CW00409667

Minerals and Mining in Antarctica

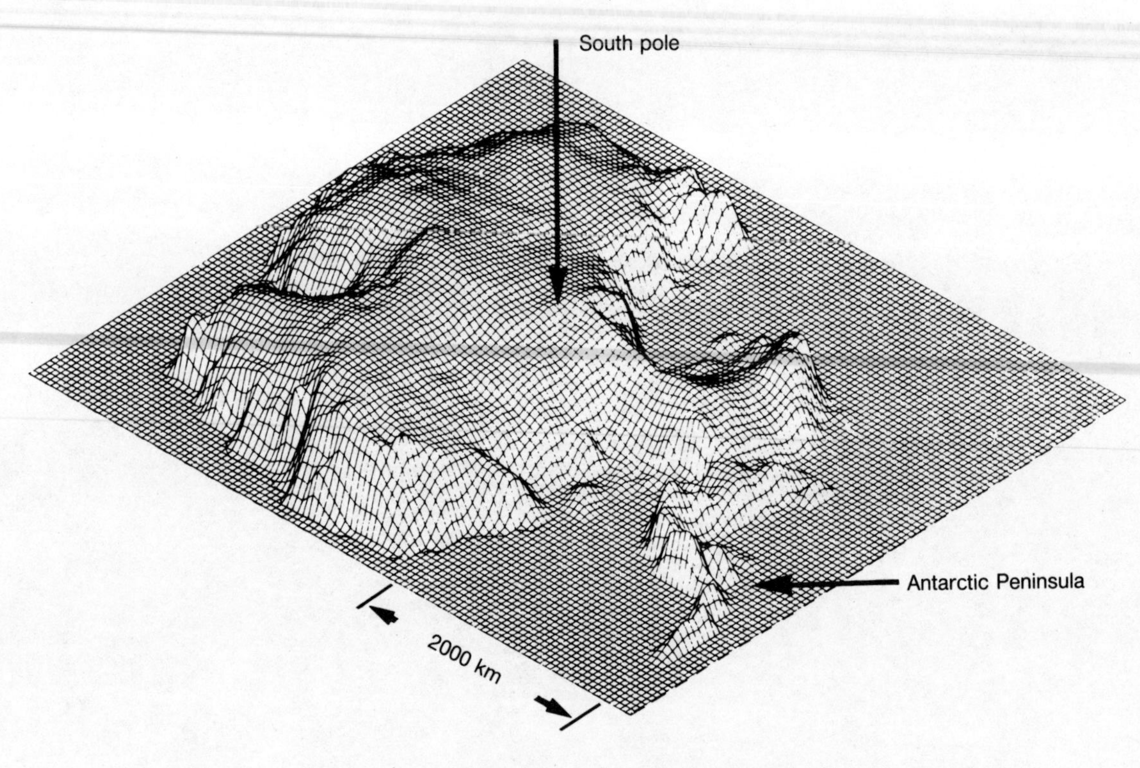

Surface of the Antarctic ice sheet as viewed from above South Georgia. This isometric view is plotted on a 50-km grid fitted to all known surface heights. Source: Drewry (1983).

Minerals and Mining in Antarctica

Science and Technology, Economics and Politics

MAARTEN J. DE WIT

CLARENDON PRESS · OXFORD 1985

Oxford University Press, Walton Street, Oxford OX2 6DP
Oxford New York Toronto
Delhi Bombay Calcutta Madras Karachi
Kuala Lumpur Singapore Hong Kong Tokyo
Nairobi Dar es Salaam Cape Town
Melbourne Auckland
and associated companies in
Beirut Berlin Ibadan Nicosia

Oxford is a trade mark of Oxford University Press

Published in the United States
by Oxford University Press, New York

British Library Cataloguing in Publication Data
Wit, Maarten J. de
Minerals and mining in Antarctica: science and technology, economics and politics.
1. Mineral industries—Antarctic Regions
I. Title
333.8′5′09989 HD9506.A58/
ISBN 0–19–854477–4

Library of Congress Cataloging in Publication Data
De Wit, Maarten J.
Minerals and mining in Antarctica.
Bibliography: p.
Includes index.
1. Mines and mineral resources—Antarctic regions.
I. Title.
TN126.D49 1985 338.2′0998′9 85–15423
ISBN 0–19–854477–4

Set by Cotswold Typesetting Ltd, Cheltenham
Printed in Great Britain by St Edmundsbury Press, Bury St Edmunds, Suffolk

For Aberra Aguma and the late Bill Morton in appreciation of their lessons in the morals of equity and needs

PREFACE

Until recently, Antarctica was a continent entirely devoid of human activity; today, it faces an increasing onslaught of 'scientific colonialism'. Science is the means by which a small number of countries are exercising their historic claims to this continent or parts thereof, whilst a mutual agreement in the form of the Antarctic Treaty guards their long-term interests. Over the last decade, many speculations about the abundance of natural resources in Antarctica have placed this Treaty under severe stresses due to its inability to cope effectively with management of exploitation and rent sharing of these resources.

Some of the resources, such as marine proteins, are being tapped in increasing amounts by Treaty and non-Treaty bound nations. Other resources, such as on-shore mineral deposits, are, as yet, untouched. Whilst there are sound scientific reasons for believing that such deposits are present throughout Antarctica, significant concentrations have not yet been proven. Moreover, it is the consensus of opinion that even if such deposits were located, Antarctic resources could not compete effectively with those exploited elsewhere in the world. Consensus, therefore, does not foresee future mineral exploitation in Antarctica as a reality.

Against a background of mining technology developed in the Arctic over the last decade, this study aims to show that for certain precious and strategic minerals the foregoing Antarctic prognosis is based on misleading assumptions. For example, it can be specifically shown that the economic climate is ripe for serious consideration of the exploitation of Antarctica's platinum potential. However, the decision as to whether or not to extract minerals from Antarctica requires far more than mere economic guidelines; there are also a number of intertwined social and political factors that must be scrutinized.

Foremost amongst these are environmental considerations. Antarctica is a unique environment, free from endogenous industrial pollution. Fears by conservationists that the onset of mineral exploitation will mark the end of this pristine 'nature reserve' are real and legitimate, especially given today's mining ethics. Therefore, the growing demands by conservation groups for the protection of this continent from any future exploitation must be diligently examined.

On the other hand, the Antarctic protectionists must, in their turn, give serious attention to some awesome facts with grave long-term consequences for the Antarctic environment in general and the world society at large, if exploration and possibly exploitation of at least certain Antarctic minerals is not pursued. Let us examine the reasons for this.

Antarctica is a large heat sink which plays a significant role in the interaction of the fluid envelope of our planet, and as such it buffers the delicate global weather and climate balance. Cold Antarctic waters and air are known to mix continuously with those of lower latitudes, which means that pollution of the atmosphere and oceans elsewhere must eventually penetrate the Antarctic region too. In fact, scientific measurements in Antarctica on a number of industrially and militarily produced chemicals and radio-isotopes have documented this mixing process beyond doubt.

Much of the pollution in the mid and lower latitudes, specifically in the highly industrial countries, is a by-product of the combustion of fossil fuels. Some of the pollutants in question are hydrocarbons, carbon monoxide, carbon dioxide, oxides of nitrogen, and sulphur dioxide. Their principal sources are power generating plants, domestic and industrial space heating-cooling, and motor vehicles. Platinum plays a vital role in the control of these health and environmentally damaging pollutants, and it is also a metal of vast potential in the future production of cost-effective and 'clean' energy from fuel cells. Platinum can thus be expected to become an increasingly important metal in tomorrow's society. Proceeding from this assumption, it is arguable that the exploitation of Antarctic platinum may be environmentally beneficial and socially profitable. However, one could ask, Why Antarctic platinum?. Surely platinum can be mined in sufficient quantities elsewhere?

The answer would be that, unfortunately, there is the inescapable fact of heterogeneous distribution of mineral riches in the world. Platinum deposits, for example, are geographically very limited in their occurrence. South Africa has inherited about 85 per cent of the world's platinum reserves outside Antarctica. Consequently, platinum supply to the world market is monopolized. Therein lies the strategic and social value of this metal, and other resources, which cannot easily be converted into real economic terms. In the context of preventing monopoly malpractices for example, there is a strong case for stimulating the development of new platinum resources. In a similar spirit, movements whose aim is the elimination of social injustices such as, for example, the institutionalized human rights violations in South Africa, can urge for exploitation of platinum elsewhere, since it can be argued that the apartheid system is deeply rooted in South Africa's profitable mining industry. Antarctica, at present, offers the only realistic potential alternative source. Moreover, there is no *a priori* reason to believe that this principle cannot be extended to include other important minerals. The pertinence of Antarctica's mineral wealth, therefore concerns not only the extent to which economic profits can be expected to induce exploitation there, but also whether or not Antarctic exploitation can also generate fundamental global gains beyond such economic profits. The sheer size of Antarctica allows scope for such an approach. Indeed, between the mineral riches of Antarctica and those of the Deep Seas, all major mineral needs can be met in sufficient quantities to influence world markets.

Naturally, it would be most rewarding if, by merely propagating *knowledge* of Antarctica's mineral wealth *and* its exploitation economics, sufficient leverage could be gained to support optimization of resource utilization elsewhere as well as the improvement of world social health. In such a scenario, Antarctica's mineral wealth could be used passively to buffer exogenous social, political, and economic instabilities, or could even be exploited as a catalyst to prevent their development. Theory predicts that this is possible; in practice it will require political moves towards centralizing management and ownership of Antarctic mineral resources so that, like those of the Deep Sea, they can be treated as a common heritage of mankind. This would demand significant concessions on the part of the Antarctic Treaty nations, a course not often recommended because of its envisaged complications. The Antarctic Treaty comes up for potential renegotiation in 1991. This leaves little time to encourage present political discussions at the United Nations and amongst the Antarctic Treaty nations to evolve into a more open dialogue, so that this colossal challenge to find an optimal solution can be faced in the most efficient way. In this intervening period, Antarctic scientists must also recognize their responsibilities and moral obligations to do their utmost towards quantifying, in the best possible way, the mineral resource potential of this vast continent. Without such a data base, political decisions with disastrous consequences for this continent and indeed for the global community, are likely to be made, leading surely to condemnation of Antarctic 'scientific colonialism' by future generations.

After exploring the concepts and spirits outlined above, this study specifically addresses some new approaches, advances, and problems related to Antarctica's mineral resource accountancy.

Johannesburg M. J. de W.
May 1984

ACKNOWLEDGEMENTS

The technical and economic skeleton for a potential Antarctic mine is based on that of the Polaris mine in the Arctic. Without the help of the Cominco geologists and management in their Vancouver, Toronto, and London offices, it would not have been possible to realistically constrain an economic feasibility study on this hypothetical Dufek Platinum Metals Mine. Initial discussions with Ted Muraro and Barry Cook in Yellowknife (N.W.T., Canada) set off a chain of stimulating and constructive conversations with them, with Mats Heimersson (chief development mining engineer at both the Black Angel and the Polaris mine), Bob Gannicott, Ross Andrews, Steve Markell, Barry Hancock, Ron Nichols, Ian Chisholm, Brian Waters, and others. My thanks to them for their frank and open discussions and help with access to the type of unpublished information which would be unheard of in most mining companies. The Antarctic community will hopefully benefit from this unusual insight.

Professor Brian Mackenzie of Queens University, Kingston, Ontario suggested to me in 1980, in Johannesburg, that the Polaris mine might be a unique example with which to evaluate Antarctic mineral exploitation. I am extremely grateful to him for introducing me into the world of mineral economics and for creating an opportunity to develop this idea further at Queens' University. The terms for a year's stay as a Queens' University Quest visiting scholar in their Geology Department were extremely generous, and financial burdens were considerably eased through the innovative approaches of their chairperson, Ed Ferrar. I am deeply grateful to him, to all members of staff in this department, and to their spouses for their understanding and help in making that year a successful foundation for this study as well as creating an opportunity for a memorable experience of Canadian living and hospitality. Additionally, I was greatly stimulated by several members of the Queens' University Economics Department, in particular through communications with John Hartwick and Nancy Olewiler during their lucid courses in the economics of natural resources.

Most of the writing was done whilst in the UK, where I was accorded hospitality both at the Geology and Mining Departments of the Imperial College of Technology, London, and the Department of Earth Sciences at Cambridge. Furthermore, in Cambridge, the Director of the British Antarctic Survey (BAS), Dr. R. M. Laws, allowed me unrestricted access to their library facilities and I am grateful for the help and guidance from many people at BAS who are much more familiar with Antarctica than I am. Discussions with Drs R. M. Laws, R. J. Adie, C. W. M. Swithinbank, and D. Limbert were useful, whilst frequent informal communications with Peter Clarkson, Janet and Mike Thompson, Chris Doake, and Christine Phillips were instrumental in the successful acquisition of Antarctic data. Gathering of Antarctic facts was boosted through the use of the library facilities of Cambridge University's Scott Polar Research Institute, a unique and unrivalled source of Polar information. Help and advice from David Drewry and A. P. R. Cooper of this Institute was extremely generous and beneficial. Hospitality in Cambridge came without any obligations: Professor Ron Oxburgh extended research facilities at the Earch Science Department without even getting me to 'sing for my supper'; Simon Lamb graciously provided me with writing and living space in his Trinity College rooms.

In London, Colin Dixon cleared the way for the use of facilities at Imperial College. Two people of this Institution were of invaluable help during this period. Firstly, Dennis Buchanan of the Geology Department was a constant inspiration, particularly when probing his knowledge of the geology of the Bushveld complex, its sulphide mineralization in general, and its platinum resources in particular. Secondly, Ralph Spencer of the Mining Department put in a lot of time and effort with helping to fine tune the economics of the Dufek Platinum Metals Mine. The Antarctic mine-models were run on his computer programme and time. Without this generous help it would not have been possible to analyse the project in such concrete terms. I also benefited from discussions with Hugh Allan and

John Stocks of the Mining Department. The latter provided ideas towards solving the water supply problem for the Antarctic mine as well as some raw-data for truck-transportation of the ore concentrate.

Antarctica is steeped in vague and nebulous politics, much of which can be misleading. Brief but clarifying discussions on this subject were had with R. Tucker Scully (Office of Oceans and Polar Affairs, US Department of State, Washington), J. Heap and A. Watt (Foreign Office, London), G. Hemmen (SCAR, Cambridge), O. A. van der Westhuysen and P. Oelofsen (CSIR, Pretoria), and Lee Kimball (IIED, International Institute for Environment and Development, Washington). However, it was undoubtedly the level-headedness and the balanced overview given by Barbara Mitchell that kept me from sliding too deeply into the insanity of this topic. Professor K. R. Simmonds (Faculty of Laws, Queen Mary College, University of London) helped to iron out some legal technicalities concerning Antarctica and the Deep Sea, whilst Keith Palmer (World Bank, Washington) explained the taxation schemes structured into exploitation projects in the Third World, as advocated by the World Bank. H. H. Vaughan (Lockheed-Georgia Co.) gave me some details on Hercules transportation aircraft; A. E. R. Budd (Johnson Matthey Chemicals Limited, UK) provided data concerning the platinum industry; R. J. McCarthy (Longyear, Canada Inc.) advised on drilling through thick ice, and G. Guthridge (NSF, Washington) supplied information on Antarctic logistics and transportation. Conversations with John Anderson (Rice University), David Elliot (Ohio State University), Ed Stump (Arizona State University), Art Ford (USGS, Menlo Park) and Peter Rowley (USGS, Denver) on specific topics of Antarctic geology and resources were particularly informative. Finally, but often most profitably, I have had innumerable discussions with innumerable friends and colleagues over innumerable beers in innumerable pubs. Too numerous to name, but too valuable not to qualify as sound sources of inspiration. I hope they all remember who they are.

The manuscript, or parts thereof, was critically read by Hugh Bergh, Sonja Begg, R. Grant Cawthorne, Peter Clarkson, David Elliot, Lynne Ferguson, Anthea Jeffery, Ken Maiden, Michael Martinson, Carolyn Mondal, Brian McKenzie, Barbara Mitchell, Walter Seelig, Kenneth Simmonds, Brian Skinner, and Peter Vale; it was types in various stages by Margaret Harrison, Fatima Variava, and Shirley Cole. Margaret Jeffery, Adele Saiet, and Rory Leahy helped with preparation of many of the diagrams. To all of them I owe my sincere thanks.

No government institutions of the Antarctic Treaty nations were interested in financing this study; neither were any reputable mining houses. I am very grateful to the Trans-Antarctic Association for a small grant to cover typing and some travel expenses whilst in the UK. A very understanding Sir Vivian Fuchs and Peter Clarkson were instrumental in obtaining this donation. Throughout the two-year period, bits and pieces were scrounged here and there. I'm grateful to my brother Michiel and many generous friends, but most of all to Lynne Ferguson who kept me fed and inspired when both savings and motivation started running out. Thanks my love.

CONTENTS

PLATES

Plates fall between pages 16 and 17.

FIGURES AND MAPS

TABLES

1 INTRODUCTION

The risk one runs in exploring a coast in these unknown and icy seas, is so great that I can be bold enough to say that no man will ever venture farther than I have done; and that the lands which may lie to the south will never be explored. (Cook 1777).

There are two fundamental differences between the north and south polar regions of our planet. Firstly, by act of nature, there is a physical difference. The Arctic essentially comprises a deep ocean basin, about six times as large as the Mediterranean Sea and similarly enclosed by continents. Its water is variably frozen over, with an ice coverage fluctuating between 5.2 million km^2 in summer months to 11.7 million km^2 in winter months. By contrast, the Antarctic constitutes a huge continental land mass, surrounded by deep oceans (Fig. 1.1). A thick ice-sheet covers this continental mass of about 14 million km^2—a surface area somewhat larger than that of Canada and Greenland combined, one and a half times that of Australia, or roughly the size of Mexico and the USA together. At the end of the southern winter (September), the sea-ice around Antarctica covers over 20 million km^2 of the southern oceans, an area larger than Antarctica itself. By the end of the southern summer (February), this sea-ice coverage has shrunk to about 20 per cent of its winter area. (Plate 1). This seasonal dynamic polar asymmetry of our planet has a profound and measurable influence on a set of interdependent global systems such as the earth's climate, variations in the planet's ocean water temperatures and circulations, its atmospheric turbulence and weather patterns, as well as on its biological activity and food sources (Washburn 1980; Gordon 1981; Knox 1983; Zwally, Comiso, Parkinson, Campbell, Carsey, and Gloersen 1983). Critical variations in some of these systems can catalyse irreversible changes affecting the surface environments of the earth (Broecker 1984).

Secondly, by act of man, there is a difference in socio-political activity in these antipodes. The Arctic region, with its indigenous Inuit (Eskimo) population on polar lands and islands, is partly nationalized, and partly shared under international jurisdiction. By contrast, the Antarctic, devoid of such an indigenous population, is neither nationalized (despite publicized claims to the contrary) nor internationally governed. These different anthropological frameworks of the Arctic and the Antarctic inspire in nations a wide spectrum of social, economic, and political values which are not easily monitored or quantified. Nevertheless, when these values are intuitively integrated with the variable physical systems of their respective polar regions, they may form a critically charged matrix which could significantly influence the well being of the global biosphere and the future welfare of mankind. As such, therefore, in both regions they warrant careful scrutiny. Here, we examine the Antarctic.

Human activity in the Antarctic is administered through the Antarctic Treaty, an extended 'gentleman's agreement' arranged amongst an elite 'club' of twelve countries in 1957, and formally sealed in 1961 for thirty years (Auburn 1982 and Appendices A1 and A2). Although seven of these countries lay claim to parts of Antarctica on historical, sectoral, and occupational grounds, under the Treaty agreement for the time being, such claims are indefinitely frozen or, as Brennan (1983*a*) prefers, kept effectively 'simmering'. With considerable success, the consultative countries to the Treaty manage Antarctica as a type of trust territory, ostensibly as a common heritage for the benefit of all mankind (Antarctic Treaty, preamble). However, their collective policies are only partly publicized. Real long term plans can only be subject to speculation since inside meetings are held in secret. The Treaty 'club' welcomes new members, but full membership with voting rights (consultative members) is allotted only to those who invest in a minimal amount of scientific investigation within the Antarctic region. Since 1961, only four countries, Poland (1977), West Germany (1981), India (1983) and Brazil (1983), have been able to afford this expensive scientific 'entrance fee', which in the case of West Germany was in the order of one hundred million dollars (Auburn 1982). Thus, because scientific activity in Antarctica dictates a monopoly on this continent (Antarctic Treaty, see Appendix A1; Neider 1980; Quigg 1983;

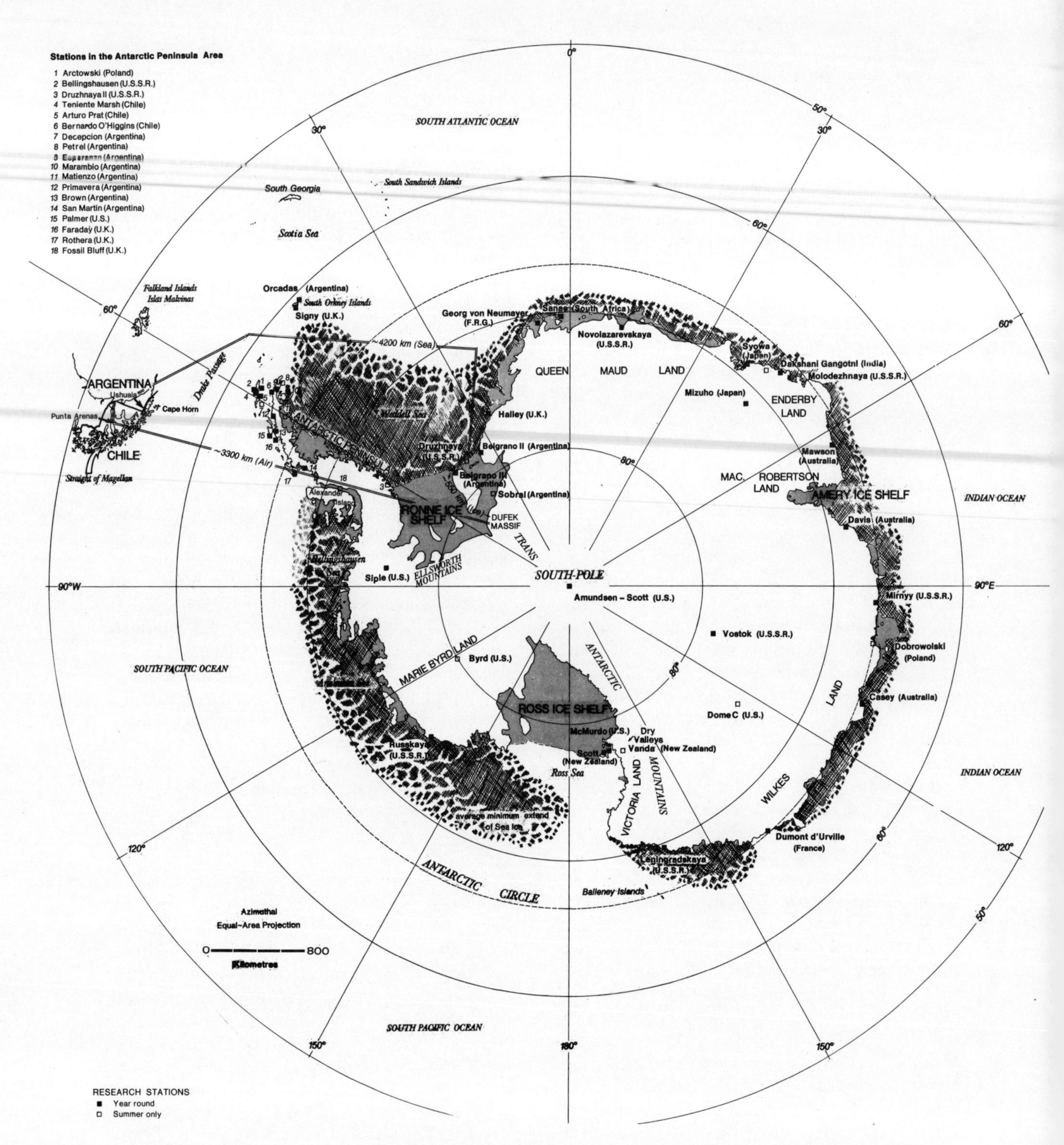

Fig. 1.1. Antarctica, showing the geographical locations referred to in the text, and the distribution of national scientific research stations. The shaded area surrounding the continent represents the average minimum extent of sea-ice. Also shown are the air, sea, and ice routes from Punta Arenas (southern Chile) to the Dufek massif, used in this study.

Nature 1983*c*), in reality the Treaty 'club' shares executive power of Antarctic affairs through means of scientific colonialism; the latter is administered through the Scientific Council for Antarctic Research (SCAR), a scientific committee of the International Council of Scientific Union (ICSU) with headquarters in England (SCAR 1981). SCAR has an annual budget of about US$ 120 000 (Kimball 1983*b*).

Over the last decades, especially since the third International Polar Year, better known as the International Geophysical Year (IGY 1957–1958, Wilson 1961), there has been a notable increase in scientific activity in the Antarctic, as evidenced by the steadily increasing output of Antarctic scientific publications (*Current Antarctic Literature and Antarctica Bibliography* 1951–1983; Fig. 1.2), and the substantial increase in monetary expenditure on the region (Fig. 1.3). Concomitant with these increases, with a predictable inherent response-lag, there is an emergence of concern for environmental alterations within the Antarctic region and by inference, concern over the potential impact of these alterations on the entire planet. The most urgent appeals for concern have been exerted by naturalist groups such as the Sierra Club (Schofield 1976), Friends of the Earth and Greenpeace (*ECO* 1983; see also Brewster 1982), Earthscan, and IIED (Institute for Environment and Development; Mitchell and Tinker 1979; Mitchell 1982, 1983; Kimball 1983*a, b*), as well as by the persistent individual efforts of concerned scientists (e.g. Roberts 1977; Lipps, in Parker 1978; Lovering and Prescott 1979; Zumberge 1979*a, b*). This concern has now been officially recognized and is being examined by the Antarctic scientific community as a whole (SCAR 1977; EAMREA 1979; AEIMEE 1981, 1983) and also within the inner political circles of some of the consultative countries to the Antarctic Treaty (US Department of State 1979, 1982; Holdgate and Tinker 1979; Todd 1983). Many individuals and organizations have recently seen their efforts of sustained 'conservation pressures' rewarded in the implementation of more rigid 'club' rules *vis à vis* the protection of plant and animal life, as well as management of renewable resources such as seals, whales, and krill (1978, entry into force of the Convention of the Conservation of Antarctic Seals; Final Act of the Conference on the Conservation of Antarctic Marine Living Resources (CCAMLR), opened for signature in 1980 and entered into force in 1982). However, the effectiveness of these measures must await further Antarctic developments and negotiations because with the expansion of scientific knowledge, familiar signs of economic entrepreneurship have emerged, and with them inevitable political overtones (Slevich 1973; CIA 1978 and updates; Mitchell and Tinker

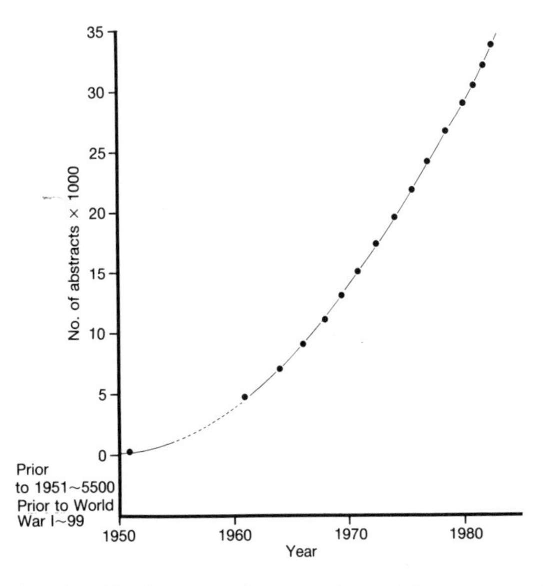

Fig. 1.2. Increase of Antarctic scientific literature from 1952 to 1983, expressed in the number of published abstracts of scientific articles quoted in *Current Antarctic Literature and Bibliography.* The numbers of publications directly related to IGY (1957–1958) are not included.

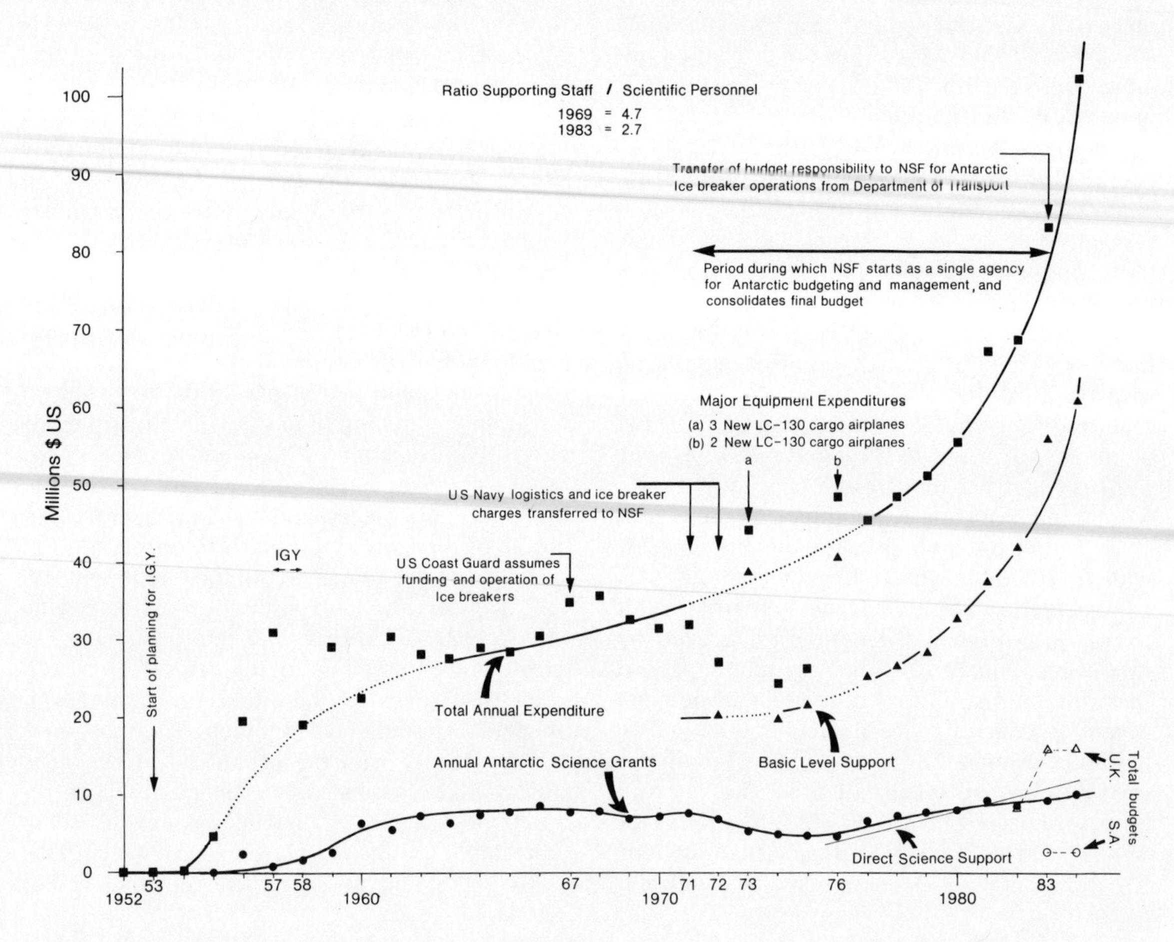

Fig. 1.3. Increases in the United States Antarctic expenditures from 1952 to 1984. This graph illustrates the 'ballooning' difference between total annual expenditures and the amounts spent on annual Antarctic science grants and direct science support. In 1984, the difference spans more than an order of magnitude, despite a decrease in the ratio of supporting staff to scientific personnel of about 45 per cent since 1969. The United States Government attributes this increasing skewness to rising fuel costs, inflation, and military pay. However, scientific expenditures show no comparable increases which might be related to rising fuel costs and inflation over the same period of time; both science grants and direct science support include substantial transportation costs to and within Antarctica. (*Science grants* are direct grants of funds to research institutions—mostly universities, some federal agencies, and others—that include sums for salaries, special research equipment, domestic travel, other direct costs, and allowable institutional overheads. *Direct science support* comprises activities in the field programme that are directly attributable to a particular science project. This includes laboratory equipment, aircraft flight-hours and international travel. *Base level support* is the cost of maintaining the US presence in Antarctica: stations, logistics, communications, new buildings, power plants, and other capital items—NSF, 1983). Expenditure fluctuations related to major equipment purchases and budgetary transfers to NSF, as indicated on the graph, have been smoothed out. The graph thus also illustrates the need to more closely examine the causes of the real increases in Basic Level Support. For comparison, the total annual Antarctic budgets of the United Kingdom (UK) and South Africa (SA) are also shown. The sudden increase of the UK Antarctic budget followed directly in the wake of the recent Falkland Islands–Islas Malvinas war. (Sources NSF (National Science Foundation) personal communications and USARP 1983.)

1979; Lovering and Prescott 1979; Auburn 1982, 1984; Quigg 1983; Mitchell 1982, 1983; Vicuña 1983; Couratier 1983; Beck 1984). Antarctica is labouring under the birth-pains of evaluating economic costs, risks, and returns. There are a number of well defined avenues for substantial economic returns via the exploitation of Antarctica. Commercial exploitation of fish, marine mammals, and increasingly krill are already well established, particularly by the Soviet, Japanese, and Polish fishing fleets. The 'club' members feel that CCAMLR provides an adequate foundation for wise management of this industry, now involving 10 nations and total catches exceeding half million tons per year (Holdgate 1983a, b; Knox 1983; Powell 1983). However, the Convention suffers from serious shortcomings for effective management (Auburn 1982). There is, for example, no adequate allowance in the 'club' formulas for (uncontrolled) entrance into this industry by non-members, and neither the subject of the economic use of these resources nor the granting of fishing permits are dealt with.

Similarly, tourism and adventurism has been responsible for more than two decades of economic exploitation in Antarctica, ostensibly in the spirit of exploration of natural environments of the 'last' kind. More than 31 000 people have visited Antarctica as paying tourists, adventurers, or as guests of scientific expeditions. Of these, more than 11 000 have visited the continent on day-flights, but the majority have travelled on ships, some of which are not registered in Treaty-bound countries (Reich 1980; Codling 1982; Auburn 1982). Although most tour operators apparently adhere to 'club' safety guidelines, tourism has already recorded its first tragedy with a heavy death toll (257) from a DC-10 air crash in 1979, and several potential ship disasters have been documented by Reich (1980) and Auburn (1982). However, no serious environmental impact studies on these issues have been conducted. Moreover, it appears from developments subsequent to these events that the Treaty 'club' shows no great interest in laying down stringent operational and safety regulations, nor are there any indications that they intend to embark upon collective enforcement of guidelines to prevent further accidents (Auburn 1984).

There is now also clear recognition by the earth-science community that oil reserves are likely to be present in the Antarctic, particularly offshore in the thick sedimentary basins of the Ross Sea and ice-shelf, the Weddell Sea, the Filchner and Ronnie ice-shelves, and the Amery ice shelf. Academic geologists have speculated on the sizes of oil reservoirs and their whereabouts, and also on the technical aspects of extraction in Antarctic waters (Splettstoesser 1976, 1978, 1979; Lovering and Prestcott 1979; Behrendt 1979, 1983; Hinz and Krause 1982; Lock 1983, Cook 1983; Gjelsvik 1983). Although present opinion favours exploitation only in the case of discovery of giant oil fields (>0.5 billion barrels), such guidelines are of course subject to a complex array of economic and political factors. Those aspiring to economic profits are probing accordingly. As part of a three-year programme (Mitchell 1983; Auburn 1984), seismic surveys with commercial overtones have been conducted in the Bellinghausen Sea (1980) and the Weddell Sea (1982) by the Japanese National Oil Corporation for the National Agency of Natural Resources and Energy, from aboard the Japanese metal mining agency's ship, Hakurei Maru. A similar survey in the Ross Sea (1983) was the last leg of this programme. Other commercial oil companies such as BP, Texas Gulf, Elf, Hunt Oil, and Total, have also shown interest in these Antarctic waters, some to the extent of trying to obtain mineral and exploration rights (Mitchell and Tinker 1979; Auburn 1982; U.S. State Department 1982; Gregory 1982; Mitchell 1982, 1983; Quigg 1983; Tucker 1983). Other commercial oil companies play influential advisory roles, for example Gulf Oil and Exxon are both represented on the U.S. State Department Advisory Committee on Antarctic Policy (Mitchell 1983; ECO 1984).

There is also a growing amount of literature published on onshore mineralization. Numerous accounts concerning the existence of Antarctica's potential exhaustible resources, including coal, uranium, base and precious metals, iron, chromium and even industrial material have been presented (Wright and Williams 1974; *Mosaic* 1978; Splettstoesser 1980, 1983, 1984; Rowley and Pride 1982; Rowley, Williams and Pride 1983; Behrendt 1983; Kameneva and Grikurov 1983; Gjelsvik 1983). Despite the speculative nature of these scientific studies, they are among the motivating factors behind the proposal by the United States, as one of the Treaty consultative parties, to develop an International Regime for Antarctic mineral resources:

'The opportunity for the United States, including United States firms, to take part in any mineral resources activities permitted in Antarctica, on a nondiscriminatory basis, is a United States objective in the development of a regime' (U.S. Department of State 1982, p. D-20).

Thus, it becomes apparent that economic and political incentives have clearly merged with the intended scientific endeavours proclaimed in the Treaty. This situation highlights questions dealing with exploration–exploitation rights and the collective environmental protection responsibilities in Antarctica. Furthermore, it focuses attention on the aspects of ownership and the legal processes of carving-up the unknown Antarctic natural resource pie amongst 'club' members as opposed to international management and rent sharing by all. In this context, a discussion is warranted which explores the speculations, controversies, and disagreements over a final regime for seabed mining, which raised similar questions during the United Nations Conference on the Law of the Sea and which culminated in the signing of the Final Act of the Convention on the Law of the Sea (UNCLOS), by 119 countries, in Jamaica in December 1982 (Lovering and Prescott 1979; de Wit 1981; Kronmiller 1980; Auburn 1982; Goldstein 1982; Archer 1983; Brown 1983; Kimball 1983, Vicũna 1983, Van der Essen 1983, Brennan 1983b). UNCLOS provides for the management of the deep sea mineral resources by an International Sea-Bed Authority in a framework of a common heritage for mankind. The experience gained from these conferences over the last two and a half decades could provide guidelines for negotiations and debates concerning Antarctic mineral resources. However, there are indications that these negotiations will have to be conducted with a greater sense of urgency than in the case with UNCLOS.

At present, there is an underlying spirit of disbelief or scepticism *vis à vis* the technical or economic feasibility of exploration, let alone exploitation, of onshore mineral wealth within the next few decades, given the remoteness of Antarctica, its harsh climate, the amount of ice-coverage (98%), the lack of suitable mining-related technology, the over-abundance of minerals elsewhere, and the potential opposition of environmental conservationists (Elliot 1976; Holdgate and Tinker 1979; Zumberge 1979 a, b, 1981; Pontecorvo 1982; US Department of State 1982; Mitchell 1982; Quigg 1983; Vicũna 1983; Holdgate 1983 a, b). This is without question the consensus of opinion of the majority of the earth scientists familiar with Antarctica, and it appears to be accepted at face value by the political representatives in the 'club', as revealed during a recent meeting in Antarctica (Holdgate 1983 a, Vicũna 1983) and a subsequent United Nations General Assembly debate (Heap 1983). One of the objectives of the present work is to examine the wisdom of this belief. Could this consensus of opinion, which may have emerged from the concept of concurrence-seeking or Groupthink (Janis 1972; Janis and Mann 1977) be unintentionally or even deliberately misleading?

For example, the Druzhnaya Base (USSR) on the Weddell Sea (Fig. 1.1) was occupied in 1975 with a cryptic announcement of intended mineral prospecting (*Antarctic* 1975), even though the Soviet Union has been reasonably frank about its intentions and interests in assessment and possible exploitation of mineral resources of Antarctica since IGY (Slevich 1968, 1973; USSR 1981; Quigg 1983). The USSR is not alone in its Antarctic ambitions: one of the three main aims of Australia's Antarctic policy is 'maintaining a scientific program which emphasises research into mineral and living resources, climate and special opportunities afforded by the region' (Australian Department of Science and Technology 1982, p. 1). Similarly, the quest for oil, gas, uranium, iron ore, and other mineral resources is part of the West German Antarctic programme (*German Tribune* 1983). Scientific investigations of the 1982–83 geological projects of Chile and of Argentina show that both deal to a significant extent (more than 25 per cent) with mineral occurrences and exploration in the Antarctic Peninsula (Argentina National report to SCAR 1981; Antarctic 1982; *Boletin Antarctico Chileno* 1982). The US National Science Foundation is funding at least one project in the Antarctic Peninsula during 1983–1984 that deals directly with Andean-type mineralization (*Antarctic Journal of the United States* 1982) and the UK will probably follow suit with a similar programme in 1984–1985. Two USA earth science projects were scheduled during the 1982–1983 austral summer to start a new potential uranium resource assessment, in the Szabo Bluff region of the Transantarctic mountains (USARP 1982, 1983; *Antarctic Journal of the United States* 1982 pp. 3,9, and 10; Zeller and Dreschhoff 1983; Dreschhoff, Zeller, Schmid, Bulta, Morency, and Tremblay 1983). A previous resource and radioactivity survey during the 1976–1977 field season conducted in the Dry Valley region found no significant uranium concentration, but reconnaissance field work in the Transantarctic mountains has located uraniferrous pegmatites (Stump, Self, Smit, and Colbert 1981; Dreschhoff *et al.*, 1983; Stump, personal communication 1982; see Fig. 1.1). All this activity points to a growing interest in Antarctic mineralization and a firm belief in the vast mineral resource potential of this continent.

In Western Antarctica, earlier US geologic and geophysical investigations delineated the Dufek igneous complex which they recognized as a vast potential 'storehouse' for economic minerals because some of the world's richest ore deposits are contained in similar igneous complexes. The Bushveld complex of South Africa, with which the Dufek complex compares favourably on geologic grounds, is the world's largest example. It has been suggested recently (Buchanan 1979*a* and *b*) that the long term future of the South African mining industry will depend to a very much greater extent on the huge resources of base and precious metals associated with the Bushveld complex. Over 30 operating mines are currently working ores genetically associated with the Bushveld complex. In 1980, sales of platinum-group metals, chromite ore, nickel, tin, fluorspar and vanadium from the Bushveld group contributed 11.1 per cent of the total value of South African mineral sales—US $ 12 500 million—*including* diamonds and gold. Sales of platinum-group metals and nickel were 74.5 per cent of total production from the complex in 1978; platinum group metals yielded nearly US $ 500 million (Buchanan 1979a, b). These metals contributed approximately US $ 860 million (1984) to South Africa's 1982 GNP. (South African Minerals Bureau, private communications 1984).

Thus, the Dufek complex is a prime potential exploitation target for ore deposits of important minerals since it is situated in a relatively accessible part of the Antarctic continent (Fig. 1.1). This has been taken into account by many, if not all, of the political representatives of the 'club' members (CIA 1978, 1981 update; House of Lords 1982; US Department of State 1982; Auburn 1982). Both the USA and the USSR have conducted recent geological and geophysical studies within and/or around this complex (*Antarctic* 1979*a*; USA 1982; USSR 1981; Gjelsvik 1983). Nevertheless, the need to revive the question of whether this complex is economically exploitable continues to be denied or avoided by geologists (e.g. Rowley *et al.* 1983; Quigg 1983). The only published quantitive evaluation of mineral exploitation in Antarctica, particularly of the Dufek Massif (Elliot 1976), and on which this negative attitude is based, is today technically and economically inadequate. Similar earlier attitudes of Antarctic geologists towards the non-viability of exploitation of potential oil reserves by the industry have already altered considerably over the last decade. Over this period rapid technical advances were made during intensive exploration for oil-fields under comparable conditions in the Arctic in the aftermath of the first oil crisis of 1973–74 (Schatz 1983). There is now more ready recognition that this, too, could become a reality in Antarctica (Roberts 1977; Auburn 1982; Gregory 1982; Holdgate 1983 *b*; see Mitchell 1983 for overview). Similarly, some of Antarctica's minerals may be exploitable on economically and technically sound grounds, given the innovative advances in Arctic exploration and mining over the last decade, particularly in Canada (Minerals Exploration NWT, 1976–1981 reports). This study proceeds with an examination of such a thesis.

The Antarctic Treaty does not deal with minerals activities in Antarctica. There is, in fact, only a voluntary restraint policy by the Treaty nations on carrying out such activities. This is binding neither on governments nor international exploration companies. The Antarctic Treaty 'club' members have therefore clearly recognized the need to establish a minerals regime for Antarctica, and serious negotiations which began in 1981 are accelerating. It is not clear, however, why there has recently been such apparent urgency in the pursuit of such a regime. Is it for political gains; is it for economic gains; is it for environmental gains, and who will stand to benefit from such gains? More pragmatically, can any future benefits justify Antarctic mineral exploitation? Is there, for example, any *a priori* reason to believe whether in today's global framework of economic, socio-political, and environmental disequilibria, the optimal solution towards a more stable international environment will not favour moves towards planned mining and management of 'strategic' minerals in Antarctica? In the spirit of the above questions, this study also examines and expands on such a hypothetical scenario.

2 A FEASIBILITY ANALYSIS OF A MINING PROJECT IN ANTARCTICA

2.1. THE ARCTIC EXPERIENCE

The possiblity of developing a hard-rock mining venture on the Antarctic mainland must be assessed within a reasonable framework of economic, technological, and geological constraints. Since much relevant factual data for such an investigation is not available from within Antarctica, this data will be generated in part through mining experiences gained under comparable conditions in the Arctic.

Despite climatic and accessibility problems which impose severe economic and technological constraints on high latitude mining, there are about 35 operating mines within the Arctic circle, 11 of which lie at latitudes above 70°N (Fig. 2.1). These mines produce a great variety of commodities including copper, nickel, cobalt, platinoids, gold, lead, zinc, iron, tin, diamonds, apatite, and coal (CIA 1978). Most of these mines are located in the Soviet Union, and are therefore difficult to assess quantitatively in terms of economics and technology.

Two mines, located in Canada (Polaris, on Little Cornwallis island at latitude 77°N) and West Greenland (Black Angel, at Marmorilik, latitude 72°N) are amenable to this type of study, and relevant data from these two mines are used in this work to illustrate the developing technology and profitability of mining in areas of permafrost and ice-cover. Both projects have incorporated innovative advances successfully to develop high latitude mining. New ideas and methods have been evolving over a period of 20 years, predominantly pioneered by Canadian and Scandinavian expertise, employed through Cominco Ltd, Canada, and Greenex A/S (a company incorporated by Cominco under the laws of Denmark, which operates the mine in Greenland; Cominco Ltd 1979, 1980, 1981;Vestgron Mines Ltd 1975, 1978–81; Fish 1974; Mikkelborg 1974; Cominco Ltd Northern Group 1979; *Mining Journal* 1981). Each of these two mines has several unique features, outlined below, which will be used in a feasibility assessment and financial analysis of a proposed mining project in the Antarctic, based on an assumed ore deposit as inferred from known, although scanty, geological data. Significantly, the two Arctic mines went into production within the last decade, during a period of growing governmental concerns for environmental deterioration related to mining (Ripley, Redmann, and Maxwell 1978). Both the Canadian and Danish governments have ensured that results of mandatory environmental impact studies at the two mine sites are stringently scrutinized. In Canada, major resource development proposals are subjected to a full social, economic, and environmental review that includes public consultations. Acceptable levels of pollution at Arctic mine sites are low by any standards; and are strictly controlled (Northern Natural Resource Development 1981; Roots 1983). There are approximately 90 items of parliamentary regulations and ordinances which affect mining operations north of 60° (Graham, McEachern, and Miller 1979; Doyle 1980). For example, the Polaris mine must comply with existing requirements such as the Northern Inland Waters Act, the Territorial Lands Act, the Arctic Waters Pollution Prevention Act, the Ocean Dumping Control Act, the Fisheries Act and others (Graham 1981). In this respect alone, these mining operations could serve as reasonable test examples for future environmental considerations in Antarctica. This will be discussed at a later stage.

Lastly, reference will be made to the Lupin gold mine recently completed in a very remote area in Canada's North West Territories, 80 km south of the Arctic circle, to illustrate a novel approach in mining logistics and transportation problems at Arctic mines away from coastal regions (Fig. 2.1). The remoteness of this mine is evidenced by the fact that the nearest town, Yellowknife, is almost 400 km away. The mine can be serviced and supplied only by air.

2.1.1 The Black Angel lead–zinc mine

This is one of the first modern mines to be brought into production under the truly Arctic

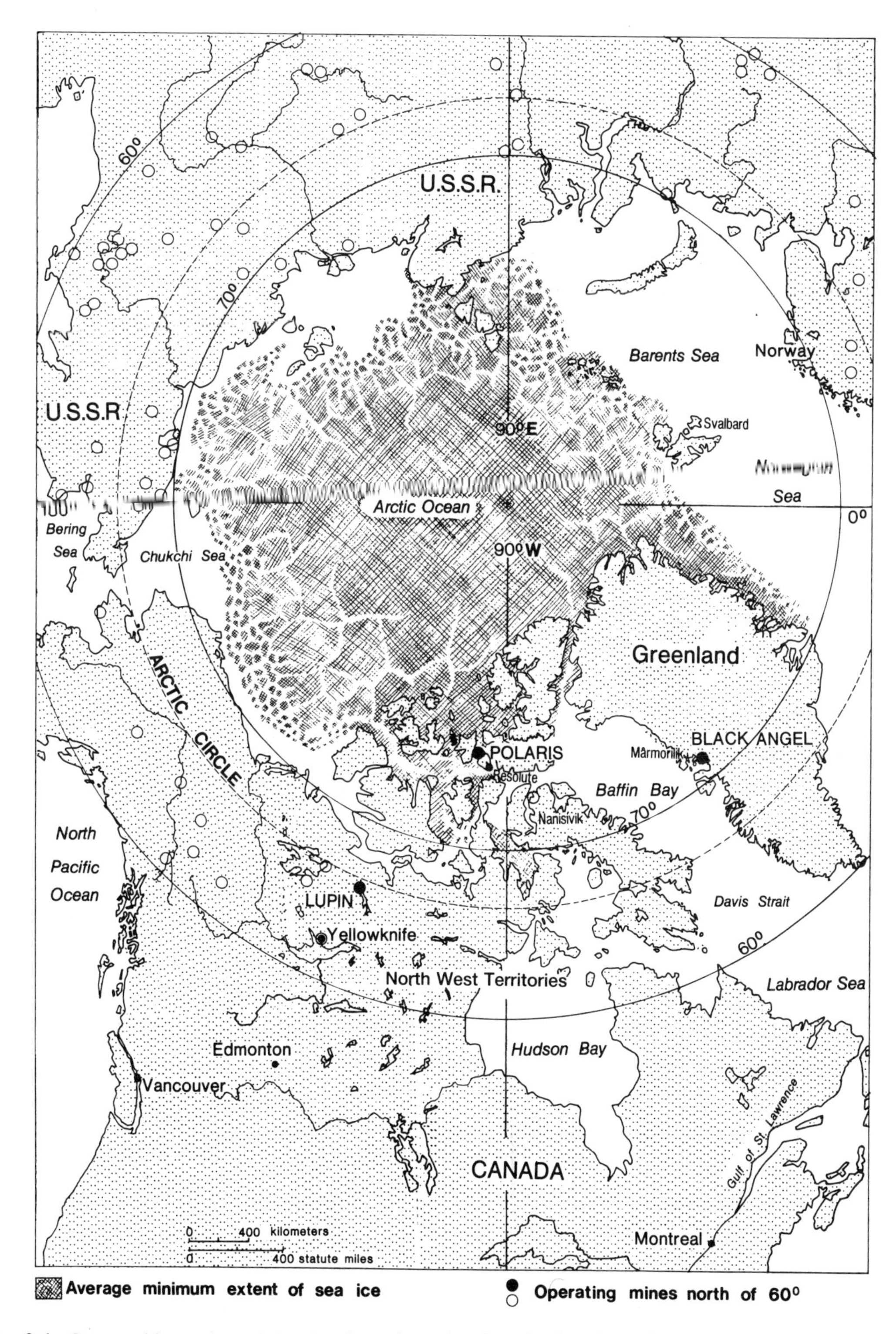

Fig. 2.1. Geographic setting of the Arctic region, showing the location of major mines north of 60°. For clarity, 10 mines on the Kola Peninsula are not shown. Mines referred to in the text are shown in black circles.

conditions prevailing in Greenland. A perpetual ice cap, up to 3 km thick, covers 85 per cent of this island. Most of the coast is dominated by mountains up to 3700 m. Access to the mine site is through a hole (incline) 600 m up a vertical cliff of permanently frozen rock, along the Agfardlikavsa fjord at Mamorilik. It is about 500 km by air (a 4 h flight) from Sondre Stromfjord, a route maintained by S-58 helicopter. From there, regular airflights operate to Copenhagen in Denmark (4 h flight time). During the construction phase, a link between Montreal and Sondre Stromfjord (approx. 3000 km) was maintained by means of a DC-6B. Twenty-eight flights were made by this aircraft to ferry personnel and smaller items of freight. The shipping season runs from June to November, an unusually long period in the Arctic, due to an offshoot of the warm Gulf Stream along part of the Davis Strait. Temperatures in the area range from a maximum of 25°C to a minimum of −35°C. Average annual precipitation is 12 cm; ground snow-loads reach 122 kg/m^2 and speeds of wind-gusts are up to 200 km/h.

Novel techniques were developed to examine thoroughly the mineral occurrence exposed along the face of the cliff with an average rock temperature of −10°C, and along which weather conditions can deteriorate rapidly. The first large-scale exploration work was carried out from June to September in 1966. This included mapping and sampling of the mineral occurrences along the cliff and geological mapping of the immediate area. The work also included the diamond drilling of six holes on the top of the cliff at 1100 m altitude. Drillers were ferried to the top by helicopter. The ore body consists of massive sphalerite and galena with a combined metal content running aroung 20 per cent. The indigenous rock is mainly white marble, part of the 1650–1790 million years old Nagssugtoquidian orogenic complex.

A more extensive programme was conducted in the summer of 1967, during which 16 diamond drill holes with an average depth of 500 m were completed. Further exploration culminated in a major programme of underground work in 1971–72. During this period, thirteen ships were used to transport 30 000 tons of construction materials, supplies and fuel from Montreal and Copenhagen. The mining project required a capital outlay of $44.1 million. Development of the mine was started in August 1973, and it was brought into full production in October 1973, on schedule and within budget. The first 35 000 ton shipment of concentrate to European smelters was made in July 1974. Ore reserves in the main or Angel zone are about 4.5 million metric tons, grading 14.9 per cent zinc, 5 per cent lead and 1 oz of silver per ton.

Several specific innovative developments stand out as major achievements in the success of this venture *vis à vis* the tight construction schedule. First was the construction by Swiss engineers and mountaineers of a cable-car system to move men and materials from the operation base at sea level across the fjord to the mine entrance, a distance of 1500 m. The personnel cableway consists of two cabins capable of accommodating 18 passengers, and with a capacity of 10 tons. The system moves 110 tons of ore per hour.

Next was the purchase of a second-hand concentrator plant, imported from Montana (USA), via Montreal by road, rail, and sea. It was modified, overhauled, and tested in Montreal, before being reassembled at the mine site. It is a conventional sulphide flotation plant, but, due to the lack of readily available fresh water at Marmorilik, sea water is used for this process. Thus, refurbishment and modification of the concentrator included considerable attention to the problems of corrosion. Design changes incorporated the extensive use of plastics and rubber diaphragm valves, with which to date there have been no serious technical problems.

The mining technique employed is the open stope room and pillar, with about 15 per cent of the ore remaining in the pillars. Because of uneven hanging wall and foot wall contacts, there is not really a typical stope dimension, but the limits of variation are 60–122 m in length, a width of 15 m, and a height of 3–21 m (see Fish 1974 for more technical details). An overcut and bench method is highly mechanized. Sea water is used in drilling. Maintenance facilities, compressors, and ventilation facilities are all installed underground; power is fed across the fjord. The mine's daily ore output is around 2000–2200 tons of ore and 200 tons of waste rock, the latter being dumped down the side of the mountain. The ore is automatically loaded into the ore skips and then carried across the fjord for treatment in the concentrator. There are 88 men working underground.

The restrictions on shipping necessitate the provision of a large enclosed concentrate storage area, and other extra facilities include large warehousing capacity and a tank-farm designed to hold sufficient fuel oil for 8 months operation.

Electric power is provided by a diesel generator with a total capacity of 7.3 MW. Fresh water for domestic and some process purposes is provided largely by desalinators which make use of waste exhaust and water-jacket heat from the generators. A prefabricated town-site has been built to accommodate about 230 people, and includes a comprehensive recreation complex and a small hospital.

Environmental control is an important element of the company's agreement with the Danish government. It requires the company to follow specific pollution control measures for the project to prevent environmental damage. Prior to the start of the operations, the waters, flora, and fauna near the mine were studied by chemists and biologists, and tests continue to be made periodically to ensure that metal elements and chemicals are within the limits set by the government. Tailings are discharged by gravity pipeline into the fjord, as approved by the Danish government. Environmental studies through oceanographic surveys by the University of British Columbia, Canada, suggest little danger of disruption of marine life in the area as the tailing fines are very similar to the large natural deposits of glacial flour on the seabed.

Since the opening of the mine, new developments have necessitated further drilling–exploration on the Marmorilik plateau, where access and transport over large distances are difficult, due to the moving and intensely crevassed ice-terrain. Rock-drilling has necessitated penetrating up to 300 m of ice coverage. Dissolved salt is used in drilling fluids to depress freezing temperatures. A salt concentration of 0.3 kg/l of water depresses freezing to about −12°C, and in this case, 900 tons of salt were needed to drill 15 000 m, using double drill-casing.

In summary, many anticipated problems connected with high latitude exploration and mining were found not to be serious at the Black Angel mine. The use of seawater in the flotation process, mining in permafrost, drilling through thick moving ice, and construction in short periods demanding tight scheduling and often under hazardous weather conditions, were all successfully achieved. Experience gained during the development of this mine provided invaluable information towards the successful construction of the next Cominco mine, Polaris, some five years later and 6° of latitude further north, conducted under much more severe conditions.

2.1.2. The Polaris lead–zinc mine

This is the world's most northerly hard-rock mine, situated on the shore of Little Cornwallis Island (77°N) and having a typical climate of very dry, short, cool summers and long cold winters. It remains dark from November through to February, and precipitation, at 13 cm of snow per year, is light, reminiscent of desert conditions. Almost continuous wind forms deep snow drifts. Temperatures vary from −50°C in winter to +15°C at the height of summer. Freezing starts in September and ice thickness varies from 0.3 m in October to 2 m in May. Permafrost extends to a depth of 430 m and summer thaw of the surface ground is minimal.

Sea access to the mine is possible only during a *six-week period* in late summer. Air transportation by DC-3 connects the mine with an airport at Resolute Bay, a small Inuit settlement about 20 min flying time to the south. From there, direct jet-transportation is maintained to Yellowknife, North-West Territories (1500 km; flying time 2–3 hours) and Montreal (4000 km; flying time 5–6 h).

Lead-zinc mineralization was discovered on Cornwallis Island in 1960; the discovery was followed up over a 10 year period by exploration activities including detailed geological mapping, geophysical surveying, and geochemical investigations, as well as initial diamond drilling. In 1972 and 1973, extensive surface drilling, development, and underground drilling confirmed a sizeable ore body with inferred reserves of 23 million tonnes of 4.3 per cent lead and 14.1 per cent zinc. Just over 3000 tons of high-grade ore were shipped to the United States for metallurgical testing and to evaluate the feasibility of shipping from the Arctic.

Initial feasibility studies in 1973 indicated that surface mine facilities would be excessively expensive. A study based on constructing the surface facilities on a ship, which was one of the alternatives considered previously for Black Angel, indicated that the concept of floating the prefabricated facilities to the site was an attractive alternative. The next 5 years were spent on engineering and environmental studies, and further economic fine-tuning, as well as on negotiations with the government over environmental standards, export agreements marketing, shipping, taxes, and overall financing for a project capable of treating 2050 tons of reserve ore per day. Capital costs were estimated at about $112.5 million with an additional working capital of $32.4 million in 1981, peaking to $35.3 million in 1985. In November 1979 Cominco Ltd announced it would proceed with the development of the mine. Initial construction material was flown in from Yellowknife, a further 13 000 tons following by summer sealift, enough for a full year's construction. The entire project, underground mine, ore-treatment plant, storage tanks, dock, and accommodation facilities for 240 people was constructed in 2 years. The first shipment of concentrate left for Europe two weeks ahead of schedule on 1 August, 1982, announced as 'an epic day for Canadian mining' (Richardson 1982). The budget exceeded the calculated costs of three years earlier by less than 1 per cent.

The key to this rapid construction was the decision to pre-build the entire mill and concentrator into a barge at Quebec and tow it 4800 km to the mine site, along the Gulf of the St Laurence and up the west coast of Greenland; stretches of ocean that are notorious for high winds and rough seas. Towing charges and insurance fees for this $55 million package amounted to $2 million. The barge arrived nine days ahead of schedule and was secured in a specially prepared site, with a crushed rock layer to form a resting base for the barge. The hull of the barge now serves as a foundation for the concentrator plant and at the same time provides storage tanks for fuel.

The mine is one of the most compact operations in Canada, with a total surface land use of 170 ha. Access to all underground workings is by 12 per cent grade ramp. Ore is crushed underground and transported 1100 m to the mill by conveyor belt. The frozen ore is competent enough to support stopes, but because it is porous and contains many ice inclusions it tends to slough when heated above 0°C. Underground temperatures are therefore kept at −12°C and ventilation systems are built with this prerequisite in mind. Due to the high Arctic temperatures during the 1983 summer, the ventilation system could not maintain the underground temperature requirement; thawing permafrost created severe ground conditions and the mine had to be closed for a time. A refrigeration plant is now being installed to prevent future problems of this sort!

The mine uses dry drillings with dust collectors. Diamond drilling uses a local recirculation brine system. Mobile equipment is designed for the Arctic environment. Power is supplied by diesel generators, capable of operating on crude oil or natural gas; total potential supply by the four generators is 9.2 MW; demand at present is 4.2 MW. Waste heat from the generators provides heat for the plant and accommodation (see Legatt 1982 for more technical details).

Much attention has been paid by various bodies (the Federal Department of the Environment and Fisheries and Oceans, the North-West Territories Water Board, the National Museum of Natural Resources, and the Canadian Arctic Resources Committee) to the impact of disposing the mine tailings, which are discharged in a nearby lake, and to the environmental impact of the mine as a whole. Investigations and studies started in 1973, six years prior to the decision to go ahead with the mine, and have been routinely carried out since by government, commercial, and university personnel. Concerns related to the influence of the mine on marine life have instigated studies on the impact of shipping traffic to and from the mine, the impact of bilge-water discharge, the possibility of oil spills resulting from use of the concentrator barge as an oil storage vessel, the adequacies of supply and sewage disposal given further expansion of mining and processing activities, and for the possible establishment of a permanent community on the site. The Inuit Association of Canada continues to probe the depth and extent of the environmental impact on wild life, and in addition, special measures are demanded for the protection of archaelogical sites on Little Cornwallis Island. A detailed chronology with a list of some of these studies is given by Graham (1981).

Ocean transportation of the concentrates to the smelters in Europe (Antwerp; approx. 7000 km) was regarded as one of the key factors which could have affected the success of the project. This movement is circumscribed by the Canadian Arctic shipping pollution prevention regulations, which specify the required ship category (e.g. ice-breaking or ice-strengthened vessels) and permitted dates of entry into the various shipping safety control zones in the Arctic. There are more than enough Canadian and European (predominantly Swedish and Belgian) commercial ice-breakers and ice-strengthened vessels of various classes and carrying capacities to compete for concentrate shipping contracts.

Another important factor realized during feasibility studies was the need to overcome obstacles of climate and isolation in attracting personnel to Polaris. Higher wages (approximately 50 per cent above that paid elsewhere in southern Canada) and quality accommodation built for comfort and with adequate recreation facilities is now believed to be one of the factors that had led to the waiting list of about 3000 job applications. Job turnover is only about 12 per cent. Miners from southern Canada work 10 weeks on and have 2 weeks off; Inuit employees work 6 weeks on and have 4 weeks off. The latter rotation is designed to meet the needs of the Inuit to continue with traditional pursuits such as hunting and trapping. Cominco pays return airfares for these vacations.

Thus in summary, underground mining in a permafrost environment, comparable to many rock exposed areas in Antarctica, has been proved to be technically possible; in the Arctic, it is being done profitably. With hindsight, it is indeed a significant understatement that technological planning, insight, and innovation were successfully executed if considered in the light of the low world commodity prices for lead and zinc when Polaris came on-stream.

2.1.3. The Lupin gold mine

This mine, developed by Echo Bay Mines of Edmonton, Alberta, is situated near the centre of the Canadian Northwest Territories, just north of the 65th Parallel, along Contwoyo Lake.

In the first full year of operation (1983), about 0.35 million tons of ore was mined, from which 3552 kg of gold was extracted (*Mining Magazine* 1983). Using a conventional cyanidation leach process, 32-kg bullion bars are cast and flown out for refining.

The weather in this region is known for fog in the summmer and extensive snow falls in the winter. Temperatures in winter fall as low as −53°C. These conditions, together with the remoteness of the mine site, presented formidable challenges for exploration and development. For example, snowstorms accompanied by strong winds, can produce an effect which makes it impossible to see the horizon (known as white-outs, which are also common in Antarctica). During construction work, such one- to two-day events were often accompanied by blocked roads, which prevented refuelling of construction heaters or the delivery of water, and also an occasional shortage of food, because aircraft could not land. The extremely cold weather also greatly increased wear and tear on the equipment. Materials become so brittle that merely hitting a pothole a little too hard results in a broken crank-case.

Lupin is an underground operation, which involved the sinking of a shaft to a depth of 357 m, establishing four mining levels and excavating an underground crushing chamber. Surface facilities include a headframe, a crusher building, a fine ore building, and a pre-engineered mineral processing plant complex to treat 1000 tons of ore per day. There is also a residence–recreation–cafeteria complex for 244 employees. The two complexes are connected by a utility corridor. A power plant, equipment maintenance bay, warehouse, and administrative offices are also housed in the plant complex. The two-storey residence block is separated from its foundation by 1.3 m high steel stilts, thus allowing air to travel under the building, preventing building heat from thawing the frozen ground, and the accumulation of snow drifts.

Before construction began in August 1980 a 1.5 km long gravel airstrip was built at the mine site. During construction, a 50-passenger Convair shuttled back and forth from Edmonton (some 1300 air km to the south-west) three times per week to carry workers and supplies. Most equipment and materials were transported on a Hercules cargo transport plane with a payload of 20 tonnes, which made an average of three flights a day. More than 1300 round trips were reportedly made. Even today's large air-cargo carriers imposed size restrictions and led to some innovative adaptions. For example, sections of the pre-engineered plant building were specially designed to be spliced together at the site.

Members of the permanent work force, primarily from western Canada, commute from Edmonton in the Convair. About 90 employees work in the surface facilities and 55 underground. They work on a rotating basis so that while 145 people are on-site at a given time, another 95 are off-site on rest period. The underground mine workers have two shifts (16 h/day total), seven days a week; they work for three weeks and have the fourth week off. Surface facility employees work in three shifts, 8 h each, seven days a week; they spend six weeks

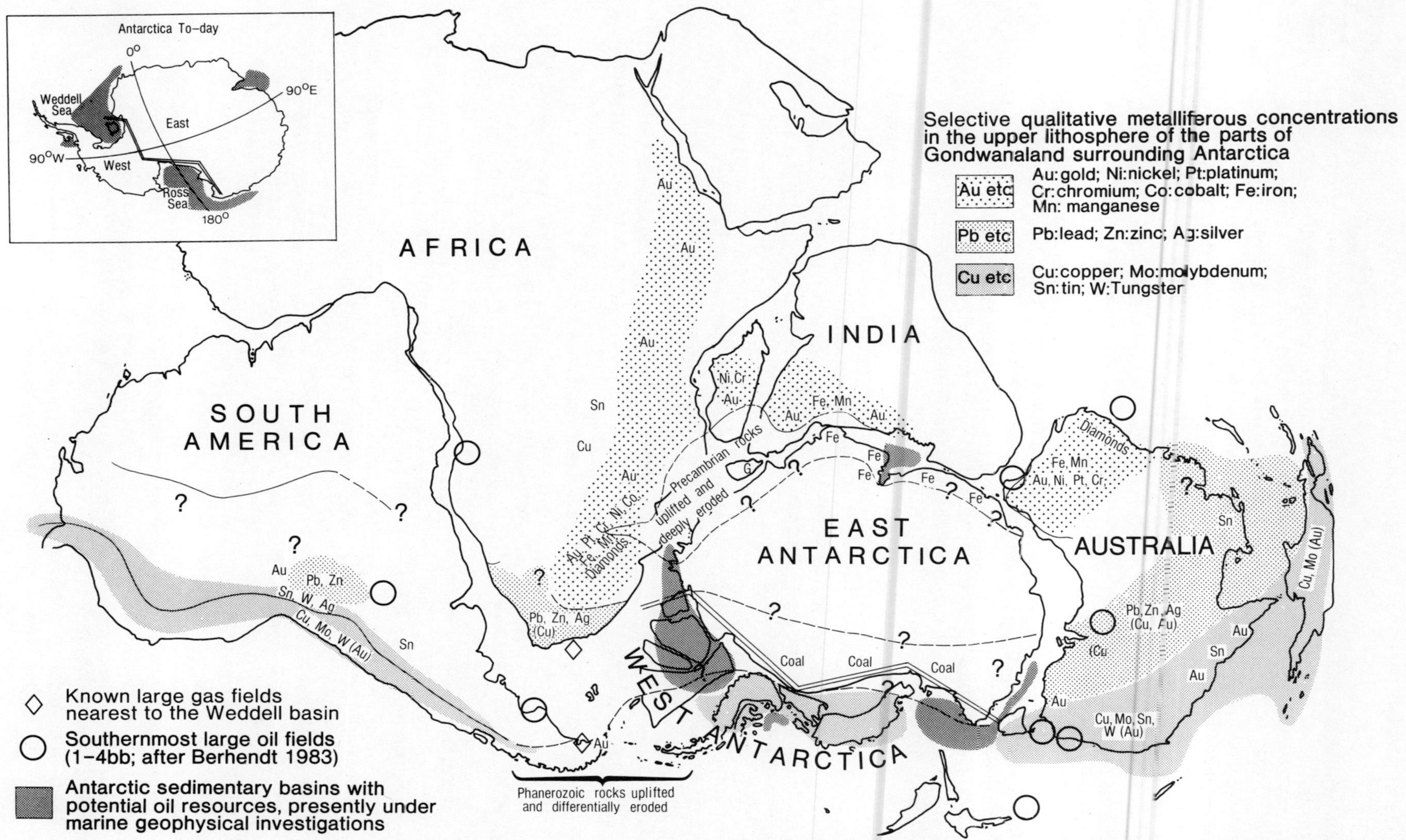

Fig. 2.2. Gondwanaland, reconstructed as it was about 180 million years ago, showing the disposition of anomalous metalliferous concentrations in selected crustal areas around Antarctica. Note gemstones—G—on Sri Lanka, which are also found in similar ancient rock formations of the immediately surrounding areas of Africa, Malagasy, India, and East Antarctica. See text for further explanation; Chapter 5 deals with Gondwanaland and Antarctic resources accountancy in detail.

Major sedimentary basins which are presently believed to have the largest Antarctic potential for oil reserves are shown both in their Gondwanaland and present-day setting (inset). There are plans to drill within the Weddell Sea section in 1986–1987, as part of the Advanced Ocean Drilling Programme (AODP), using a commercial drilling platform of the SEDCO-472 type (Elliot 1984). This work, and other marine geophysical investigation around Antarctica, is monitored by a SCAR Working Group in solid earth geophysics.

Some occurrences of coal in Antarctica are also indicated (after Splettstoesser 1980), but further discussion on the energy resources of Antarctica are beyond the scope of this book. (After de Wit *et al.* 1985.)

on-site followed by two weeks off (*Mining Magazine* 1983).

Thus in summary, underground mining for precious metals, in a permafrost environment, isolated from port and shipping facilities and without a readily available conventional infrastructure, such as might be comparable to areas in Antarctica, has already been successfully achieved.

2.2. AN ANTARCTIC EXPERIMENT

2.2.1. Exploration for a target

Apart from large coal resources along the Trans-Antarctic Mountains, and iron-ore in East Antarctica, the presence of significant mineral deposits in Antarctica has not yet been proved. Thus, any attempted technological and economic evaluation for a prospective mine in Antarctica can at best be based on circumstantial geologic evidence. Nevertheless, there is a sound scientific basis for believing that there are number of well defined areas in Antarctica that have a high probability of ore-grade mineral concentrations (Fig. 2.2; see Chapter 5). The most obvious and most commonly referred to are those that might occur in the Antarctic Peninsula of West Antarctica (Fig. 2.3). Rocks from this curvilinear mountain belt are so comparable in type and association to those that consititute the fundamental architecture of the Andes in South America, that speculations about similar ore deposits in these two belts are reasonable.

The central Andes of Chile and Peru are the 'home' of 'porphyry-type' mineralization, a name derived through its association with a characteristic suite of granitoid rocks. Porphyry copper and/or molybdenum deposits are the most common examples. They occur with or without associated gold and a host of other elements, although the latter seldom occur in significant amounts. Other deposits in the central Andes of Peru, Bolivia, and Argentia are rich in tin and

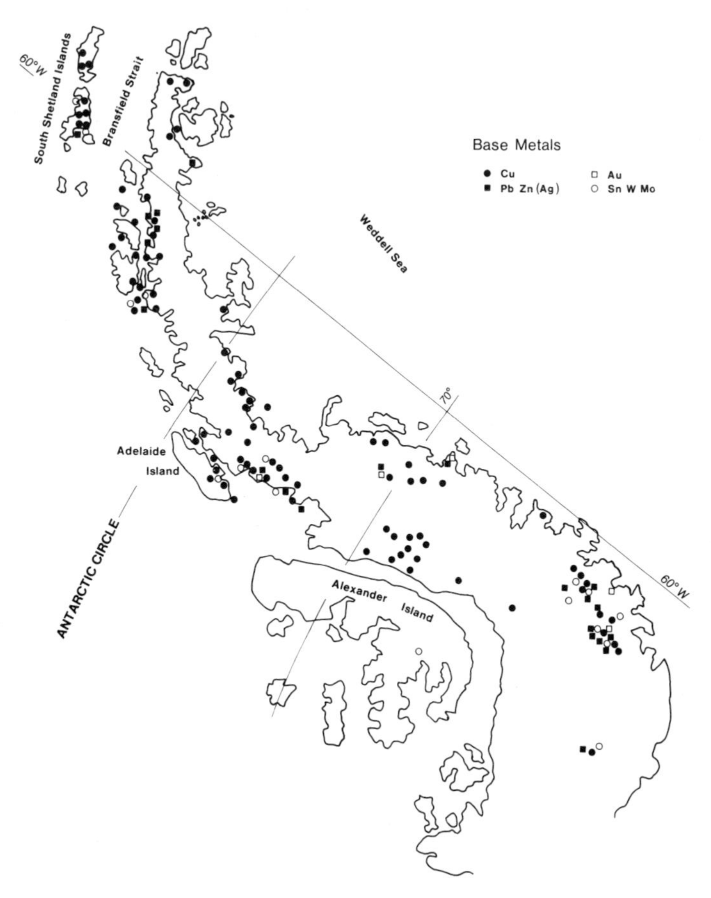

Fig. 2.3. Locations of reported mineral occurrences along the Antarctic Peninsula. (Sources: British Antarctic Survey (1979–1983) and J. Thomson (B.A.S., personal communications, 1983).)

tungsten, often associated with silver and other minor elements.

Without question the overall rock relationships within the Central Andes and Antarctic Peninsula are very similar. However, there are also notable differences which may be equally important *vis à vis* the presence or absence of mineral deposits of economic significance. What is known is that within the Antarctic peninsula mineral occurrences of copper and molybdenum, occasionally with gold, lead and zinc, have been observed (Wright and Williams 1974; British Antarctic Survey 1979–1983: Rowley *et al.* 1983; Rowley and Pride 1982; Vieira, Alarcóin, Ambrus, and Olcay 1982; Hawkes and Littlefair 1982, Pride and Moody 1982; Fig. 2.3). Indeed, mineral occurrences are so abundant that elsewhere in the world, even in the high Canadian Arctic or Greenland, such an area would be intensely prospected.

Nevertheless, a major drawback to the challenge of further exploration in this region is that most porphyry-type deposits are of low grade ore, dispersed through vast rock volumes. Consequently, they must be mined by methods which may require opencast mining: large scale rock excavations which create gigantic holes that scar the landscape. In Antarctica this would require a technology in terms of mining and environmental procedures that have as yet no counterparts in the Arctic for comparative economic modelling. Thus, despite the fact that in terms of logistics the Peninsula provides some of the most accessible areas in Antarctica, it is not considered any further in this part of the study. It must be stressed, however, that this dismissal of porphyry mining potential in the Antarctic Peninsula may be overdone. There are some world-class **Andean underground** porphyry mines (e.g. El Teniente, Andina, El Salvador, San Manuel). Furthermore, grades in some cases are high enough to make these producers among the lowest in cost in the world. Thus, they could have the ability to absorb high remoteness costs.

Along the opposite coastline of the Antarctic continent, recognition of very old rocks of the Archaean era, greater than 2500 million years in age, has now been clearly demonstrated. Similar rocks make up the fundamental architecture of large area, known as cratons, of all continents, although their geological history is at least an order of magnitude less well understood as compared to that of rocks which constitute much younger terrains such as the Antarctic Peninsula.

The cratonic rocks of Antartica which constitute much of East Antarctica have been most closely investigated by Soviet and Australian scientist, with significant United States, South African, Norwegian, German, and Japanese contributions. As from 1982, Indian investigators have also started research activities and their observations may be expected to yield key results since there appears to be great similarity between parts of the craton of Antarctica and that of India (Fig. 2.2).

Significantly, the Archaean cratons of the continents of the southern hemisphere are hosts to many small but rich and profitable gold deposits, diamond-bearing kimberlite and related rocks, as well as to substantial nickel occurrences. For example, the craton of south-western Australia contains 14.5 per cent of the world's total identified resources of nickel in sulphide ores, with grades greater than 0.8 per cent nickel (Groves, Lesher, and Gee 1984) Sixty-five per cent of the 1982 world diamond production was from the southern African and Australian cratons (Mining Annual Review 1983) and the figure is projected to exceed 75 per cent in 1985 (*Lyons* 1983). Similarly, gold from these terrains is a significant source of income in countries such as South Africa, Zimbabwe, India, and Australia. There are sound geological reasons for believing that the Antarctic craton should contain similar riches.

Significantly, on the continents of the southern hemisphere gold has been mined over lengthy periods. The majority of the presently known deposits were historically exploited (in East Africa, many deposits were mined in prebiblical times) for their rich, oxidized sections with 'free or native gold'. It is only relatively recently, since the introduction of the cyanidation process in the late nineteenth century, that it has become possible to 'liberate' gold chemically from the reduced but much less concentrated parts of these gold deposits. Most primary gold production has been recovered by this method for decades, and today these 'poorer sections' sustain these mines. It is safe to assume that the rich oxidized 'eyes' of similar gold deposits in Antarctica remain to be found. At present day gold prices these bonanzas would increase the likelihood of profitable exploitation of small gold deposits.

Accessibility in this part of Antarctica is now well established through the presence of permanent Soviet, Australian, Japanese, and South African scientific bases along the coast. The

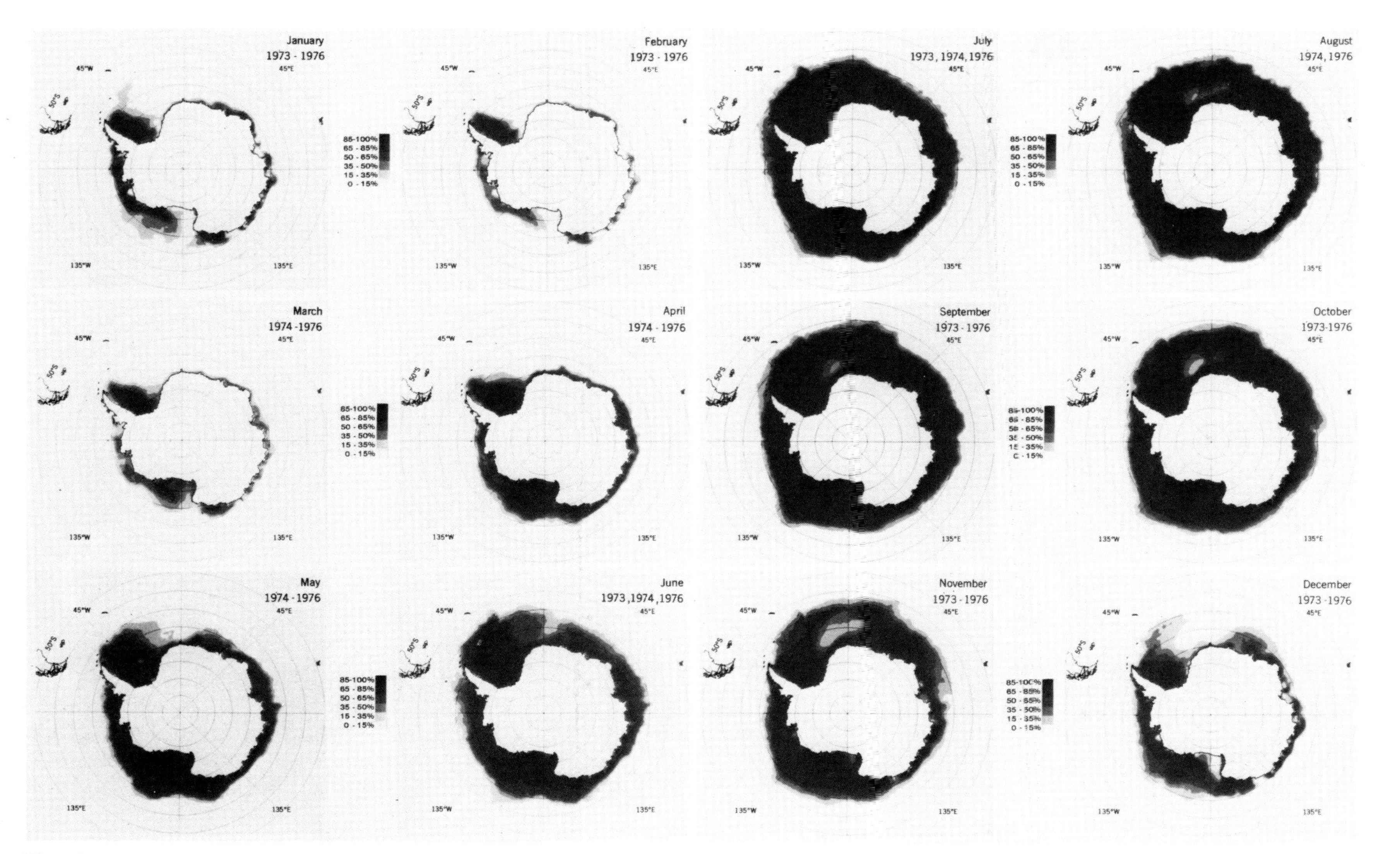

Plate 1. Mean monthly ice concentrations around Antarctica between 1973 and 1976, averaged for the years indicated. These diagrams vividly depict the dynamics of the growth and decay of the Antarctic sea ice, which, each year, recedes from a maximum area of approximately 20×10^6 km^2 in September to a minimum area of approximately 4×10^6 km^2. Data obtained from the electrically scanning microwave radiometer on the Nimbus 5 satellite, for 41 months of the 4-year period (from Zwally, Comiso, Parkinson, Campbell, Carsey, and Gloersen 1983).

Plate 2. Black Angel lead–zinc mine, Marmorilik (71°N), Greenland, in summer time. View from the head of the concentrator up to the Black Angel mine entrance (arrow), 600 m above sea level in the fjord. The namesake of the mine is clearly depicted on the cliff above and to the right of the portal. The ore buckets (foreground) move 110 tonnes of ore/h. Photograph courtesy of Cominco Ltd.

Plate 3. Late summer view from the Black Angel mine entrance, southerly on to the concentrator and townsite, with shipping active in the partly frozen over Agfardlikavsa fjord. Photograph courtesy of Cominco Ltd.

Plate 4. Polaris lead–zinc mine, Little Cornwallis Island (77°N), North West Territories, Canada. View easterly at the Polaris site. Left middle ground: the loading pier and the concentrator plant. Right middle: concentrate storage. Background: accommodation site. Photograph courtesy of Cominco Ltd

Plate 5. Pre-built barge, housing the 2000 tonnes-per-day mill, concentrator, and ancillary processing facilities, in the Davis Strait *en route* to Polaris, mid-August 1981. Photograph courtesy of Cominco Ltd.

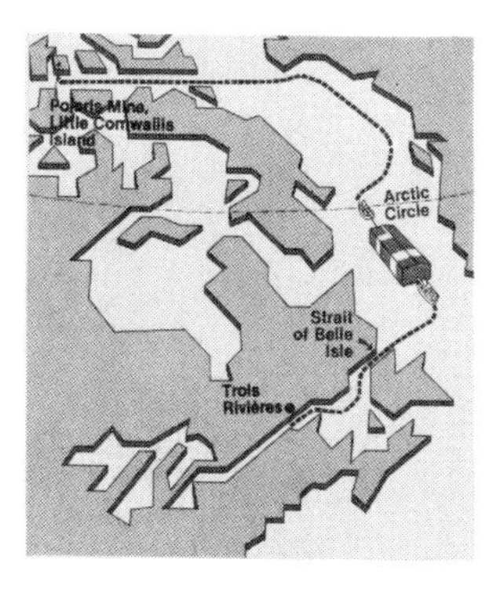

Plate 6. Near-arrival at Polaris of the concentrator plant, following the 2-week, 4800 km journey from Trois Rivières, near Montreal, along the Gulf of St. Laurence and up the west coast of Greenland (left), summer 1981. Photograph courtesy of Cominco Ltd.

Plate 7. Typical coastal outcrop along the Antarctic Peninsula, between the South Shetland Islands and Marguerite Bay. The rock exposures display a continuous cross-section through the Antarctic batholith and its volcanic cover at different crustal levels. Porphyry-copper and/or molybdenum deposits are likely to be present along parts of the Peninsula, where abundant mineral occurrences of copper and molybdenum, occasionally with gold, lead, and zinc have been reported—see Fig. 2.3. Photographs by author, 1975.

Plate 8. Mountains of the Jurassic Dufek layered complex, one of the world's largest known mafic igneous intrusions (about 50 000 km^2). Note the near horizontal igneous layering. The complex was discovered in 1957 but has subsequently only been sparingly studied. It is nevertheless recognized by earth scientists and by the Antarctic Treaty nations' politicians as a potential 'storing house' for important economic and strategic minerals: the complex is one of Antarctica's most favourable targets for exploration of minerals such as platinum-group elements, nickel, chrome, and vanadium, amongst others. The complex is used in this study for a feasibility analysis of a hypothetical platinum mine, to evaluate the costs and benefits of mining in Antarctica. Photograph courtesy of J. C. Berhendt, US Geological Survey.

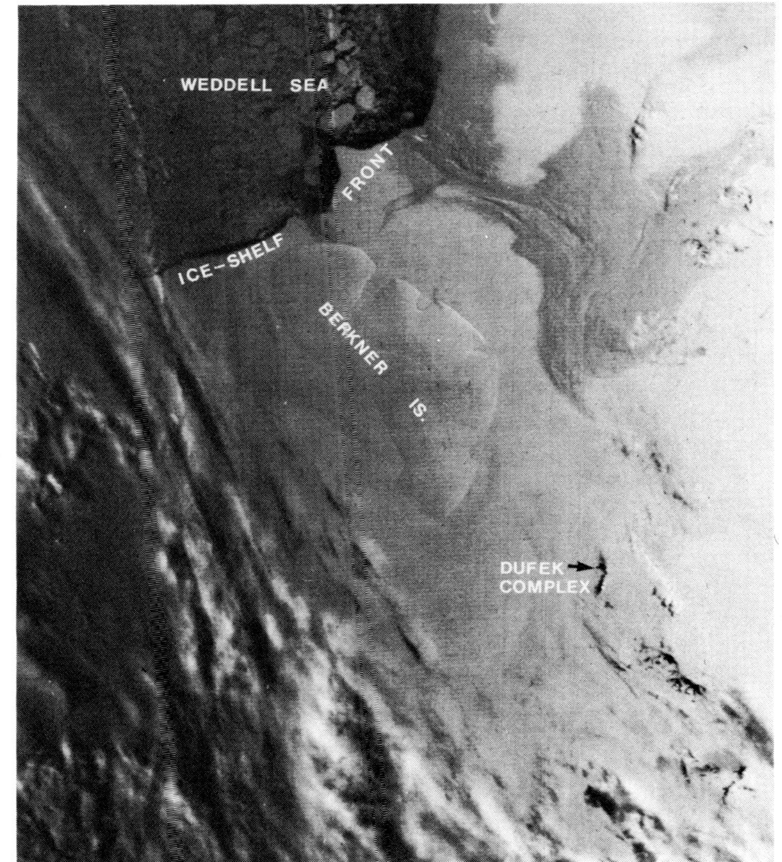

Plate 9. Satellite picture (NOAA6, southern hemisphere, 30 December 1980) of the general area depicted in Fig. 2.10, showing the setting of the Dufek igneous complex, about 500 km south of the ice-shelf front of the Weddell Sea. The sea-ice of the Weddell Sea is extensively broken up. The lower right-hand part of the area is covered by cloud formations.

Plate 10. Physiographic sketch of the north-face of the Dufek complex, viewed south from Cairn Ridge, one of the potential localities chosen for the hypothetical Dufek Platinum Metals Mine. Shown are some of the localities and features named in the text and indicated on Fig. 2.4. Photographs are close-ups of the summits of Aughenbaugh and (aerial view) Neuberg Peaks—AP and NP respectively—as shown on the sketch. The type section of the Aughenbaugh Gabbro is along the left centre spur of the right-hand photograph. Horizontal distance seen on the sketch is about 18 km. FPM, Frost Pyroxenite Member of Aughenbaugh Gabbro (dark layers on AP and NP); NPM, Neuberg Pyroxenite Member; WP, Walker Peak. Sketch by Tau Pho Alpha; photographs courtesy of A. B. Ford, US Geological Survey.

Plate 11. Physiographic sketch of the western part of the Dufek igneous complex, viewed from Hannah Peak. Horizontal distance is about 10 km. The photograph shows an aerial view south of Walker Peak (WP, extreme left on sketch), with the paler-coloured layered cumulates of the Walker Anorthosite, about 200 m below the thin dark layer of the Neuberg Pyroxenite Member (NPM). Arrow marks the contact between the Walker Anorthosite (WA) and the Aughenbaugh Gabbro (AG). AP: Aughenbaugh Peak; NP: Neuberg Peak. The valley in the foreground of the sketch might be considered as a potential locality for the hypothetical Dufek Platinum Metals Mine. Sketch courtesy of Tau Rho Alpha; photograph by A. B. Ford.

Plate 12. Overland transport near Sanae—the South Africa scientific base. Similar conditions prevail between the ice-shelf front of the Weddell Sea and the Dufek complex. This will allow for the building and maintenance of an ice-road at relative low costs. Crevasses can be crossed using ice bridges. Hovercraft transport may prove to be the most cost-effective for future projects which require long overland hauls across such relatively flat terrain. Photograph courtesy of H. Bergh.

Plate 13. Bay-ice/shelf-ice boundary, close to Sanae, summer 1983. The ice-cliff of the shelf is about 20 m high. Water depth below the ice-shelf in this region is about 450 m. The bay-ice in the foreground is up to 2 m thick and breaks up during the summer season, allowing ships to moor along the edge of the ice-shelf to facilitate off-loading of scientific and other equipment. Similar conditions prevail along the Weddell Sea ice-front (see Plate 9 and text). Photograph courtesy of H. Bergh.

Plate 14. Catabatic cloud formations over west side of central Scott Glacier, Trans-Antarctic Mountains, due to gravity driven winds. These winds originate over the central, colder interior of the 2000 m Antarctic Plateau, from which the cold heavy air is driven outward towards the lower coastal regions. The Pensacola Mountain Ranges partially shelter the proposed Dufek Platinum Metals Mine from these harsh weather conditions. Photograph courtesy of E. Stump.

Plate 15. Geological reconnaissance investigations along the Antarctic Peninsula, using the US research vessel *Hero*, January 1975. This picture displays the excellent rock outcrops, along the coastline, where exposed; accessibility and logistics support, however, pose major constraints on the amount and quality (detail) of work that can be carried out by geologists during such investigations. Photograph, author.

Russians in particular have accumulated a wealth of experience with large-scale overland operations to Soviet bases such as Vostok, some 1000 km inland (Fig. 1.1). Thus, logistically and geologically it is sensible to regard parts of the East Antarctic cratonic areas as mineral exploration targets. However, since no published data is available to evaluate signatures of potential mineral deposits, such as gold, in sufficient detail, this area, like the Antarctic Peninsula, is not suitable for further consideration in this part of the study. Nevertheless, it should not be too difficult to tailor a Lupin-type gold operation to a small gold occurrence in Antarctica, when found. Should a prospect turn out to be in a reasonably accessible region, such deposits may well prove to be economically viable propositions.

Along the opposite margin of East Antarctica lies a possibly mineral-rich belt which connects the mid-Proterozoic (2000–1000 million years old) Pb–Zn–Ag-rich belts of Australia and southern Africa (Fig 2.2). However, very little is known about the geology of this region of Antarctica, due mainly to its inaccessibility and extremely limited rock exposures. Therefore, despite the fact that this belt hosts some of the world's largest lead–zinc–silver mines in Australia and South Africa, its Antarctic potential will not be discussed further in this chapter.

One of the other potential areas for mineralization constitutes a zone that straddles the central mountain chain, the Transantarctic Mountains, which delineates the enigmatic boundary between the two vastly different geological provinces of East and West Antarctica (Figs 1.1 and 2.2). This zone, manifested by a concentration of dark igneous rocks known as the Ferrar dolerites and Kirkpatrick basalts, terminates at its northern extremity in a large geologic anomaly, known as the Dufek Massif. This massif, of Jurassic age, constitutes a differentiated layered mafic igneous intrusive complex that occupies nearly the entire northern part of the Pensacola Mountains (Figs 2.4 and 2.10). The Dufek intrusion was discovered in 1957 (IGY) during geographic and geophysical exploration of the inner reaches of the Filchner and Ronne Ice shelves (Neuberg *et al.* 1959). It has subsequently been studied independently by American and Russian geologists and geophysicists over a period of at least 18 years (Huffman and Schmidt 1966; Schmidt and Ford 1969; Ford, Carlson, Czamanske, Nelson and Nutt 1977; Behrendt, Henderson, Meister, and Rambo 1974, Behrendt, Drewry, Jankowski, and Grim 1980; USA Report 24 1983; USSR 1981; German Democratic Republic: Antarctic Research Activities 1982) and its mineral potential has been a topic of speculation both in the USA (Wright and Williams 1974; Ford 1976, 1983; Rowley *et al.* 1983) and in the USSR (Slevich 1968, 1973; *Antarctic* 1975; Auburn 1982).

The recognition of this complex as a 'storing-house' for important economic minerals is largely conjectural. It is based on geological and mineralogical comparisons of collected rock samples to similar rocks that occur in large layered intrusions elsewhere, and whose immediate relations to important mineral bearing layers within these intrusions are well known. In broad outline, mafic layered intrusions, such as the Dufek, are thought to represent gigantic fossilized reservoirs, possibly in excess of half a million km^3, of hot liquids (magma) which emanated from the mantle below the Earth's crust. Following ascent into the upper parts of the crust, at temperatures around 1100°C, the liquids, trapped as 'intruders', cooled and crystallized into a host of mineral phases, which were differentially precipitated within the reservoir in layers according to fundamental laws of physical chemistry and fluid dynamics. The precise sequence of liquid ascent and subsequent precipitation of minerals from this magma is poorly understood, but the overall geological, mineralogical, and chemical similarities of these vast intrusions point to a common mechanism. Thus, given a specific mantle source region for the liquid and a similar crustal host environment for its ascent and final residence, similar ore precipitates ought to be present at predictable levels in all large intrusions. For example, chromite layers, with associated platinum are confined to the lower parts of these complexes, as are sulphides of nickel, copper, platinoids, and cobalt; whilst iron with associated vanadium occurs at higher levels. The relative location of the economically valuable minerals can be further 'fine-tuned' given more detailed knowledge of the mineralogical and chemical variations throughout the layered sequences as a whole, and the application of this technique has therefore a great bearing on reliable prediction and location of such ore deposits. For example, impetus to the successful exploration of platinum mineralization in the layered Stillwater intrusion in Montana, USA, was initially modelled on such a comparison of its geological,

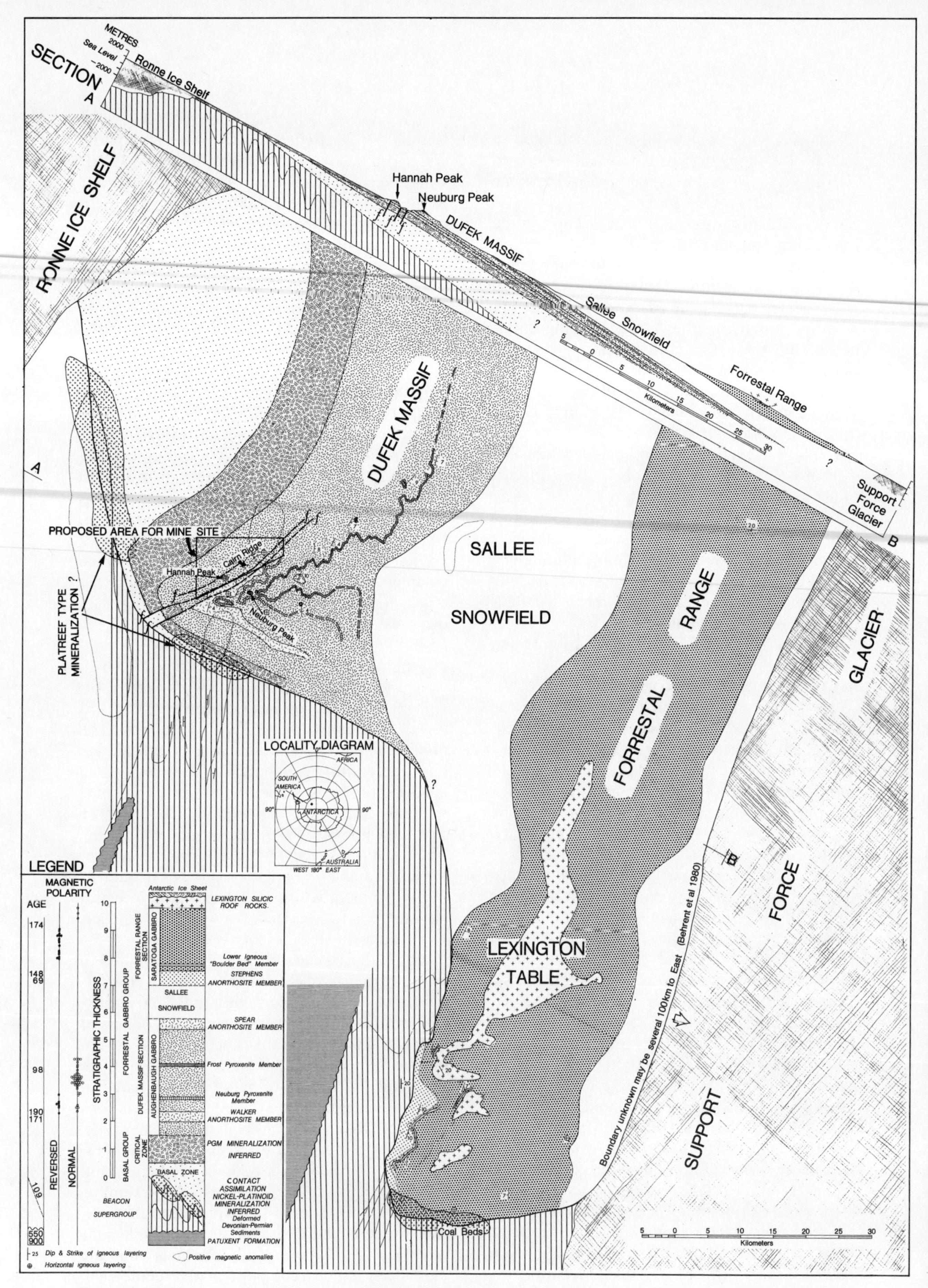

Fig. 2.4. Geological map and section of the southern part of the Dufek igneous complex, Antarctica. Also shown is the proposed area for the Dufek Platinum Metals Mine site, and the areas of possible contact-assimilation related mineralization of the Platreef-type, found within the Bushveld igneous complex. The magnetic polarities shown in the legend are taken from Beck (1972). Figure 2.10 shows the Dufek complex in a more regional framework. (Modified after Ford *et al.* 1978a, b.)

mineralogical, and chemical similarities with those of the mineral-rich Bushveld layered complex in South Africa (Conn 1979).

2.2.2. Comparative geology of the Dufek, Bushveld, and Stillwater complexes

The Dufek complex falls without doubt in a class of mega-intrusions of which only two are known to exist. A combined aeromagnetic and radio-echo ice-sounding survey over the Dufek intrusion suggests a *minimum* area for the complex of about 50 000 km^2 (Behrendt *et al.* 1980; Figs 2.4 and 2.10). This compares with 66 000 km^2 reported for the Bushveld complex, the biggest yet identified on earth (Fig. 2.5). Additionally, the presence of large amplitude magnetic anomalies to the north and east of the surveyed area suggests that the Dufek intrusion has still greater extent (Behrendt *et al.* 1980). The complex is estimated to be 8–9 km in thickness (Ford 1976). Geological field work and laboratory investigations of systematic sample collections from practically all exposed parts of the intrusion (Ford and Boyd 1968; Schmidt and Ford 1969; Himmelberg and Ford 1976; Ford 1970, 1976) have confirmed and further elaborated on the earlier suggestions of Aughenbaugh (1961) and Walker (1961) that the intrusion is similar to major stratiform intrusions elsewhere in the world such as the Bushveld and Stillwater complexes. Although the Bushveld complex is more than an order or magnitude larger in size than the Stillwater complex, important concentrations of platinum group elements or metals (PGE; PGM) occur at similar, well-defined stratigraphic horizons in both these intrusions. These ore-bearing horizons are known as the Merensky and the H.P. reefs, respectively (Figs 2.6 and 2.7).

The stratigraphy of the Dufek intrusion, described by Ford (1976), is diagramatically illustrated in Figs 2.4 and 2.6. This shows that the layered cumulates are exposed in two partial, non-overlapping stratigraphic sections. The lowermost 1.8 km of exposed rocks makes up the Dufek Massif section while the upper exposed section forms the Forestal Range. The combined thickness of these two sections is considered to be of the order of 6 to 7 km and the geophysical evidence suggests that the unexposed basal portion is 1.8 to 3.5 km thick. The total thickness of up to 9 km of mafic rocks is, therefore, comparable to the thickness of the Bushveld complex (also shown in Fig. 2.6). In further comparison with the Bushveld and Stillwater complexes, known mineral resources of chromite, platinum-group metals, nickel and minor cobalt, copper and gold should be confined to the lower concealed basal section of the Duflex complex.

Similarly, iron (as magnetite) with vanadium should then be enriched in the upper, partly exposed regions. Evidence for this has indeed been reported (Wright and Williams 1974; Himmelberg and Ford 1977; Ford 1983). For example, iron occurs in the upper parts as magnetite concentrations of 70 per cent to 80 per cent in layers as much as several metres thick. Thinner layers of up to 100 per cent magnetite have also been recorded. These magnetite-rich layers have in places measured vanadium contents of up to 0.1 per cent, although this is an order of magnitude less than that measured in the Bushveld. Field exposure is insufficient to establish the presence of other expected resources. It is possible, however, to obtain an insight into the nature of unexposed rocks from a comparison of the petrography and chemistry of the mineral phases of the two intrusions. This is because sequential crystallization from magma is known to lead to systematic trends in mineral compositions. The phase chemistry of pyroxenes has been shown to be a particularly reliable indicator of mineral fractionation following intrusion of a liquid, allowing accurate stratigraphic correlation between different complexes, since pyroxenes are virtually ubiquitous throughout the mafic igneous rocks.

Pyroxenes from layered intrusions fall into two groups, the Ca-rich and Ca-poor varieties. Where both types co-exist in a sample, the tie-lines which join them have a precise orientation which is the same for all natural rocks in which the constituent phases crystallize in equilibrium with one another. First formed pyroxenes will be magnesium-rich while the last pyroxenes to crystallize will be iron-rich. This chemical fractionation trend can be correlated directly with stratigraphic height, the lowest rocks in the sequence having the most magnesium-rich pyroxene pairs. Conversely, samples from the top of the sequence are very iron-rich.

Thus, in the absence of drill core data, identification of potential mineralization in the lower parts of the Dufek massif can be attempted by comparing the known geological, mineralogical,

(a)

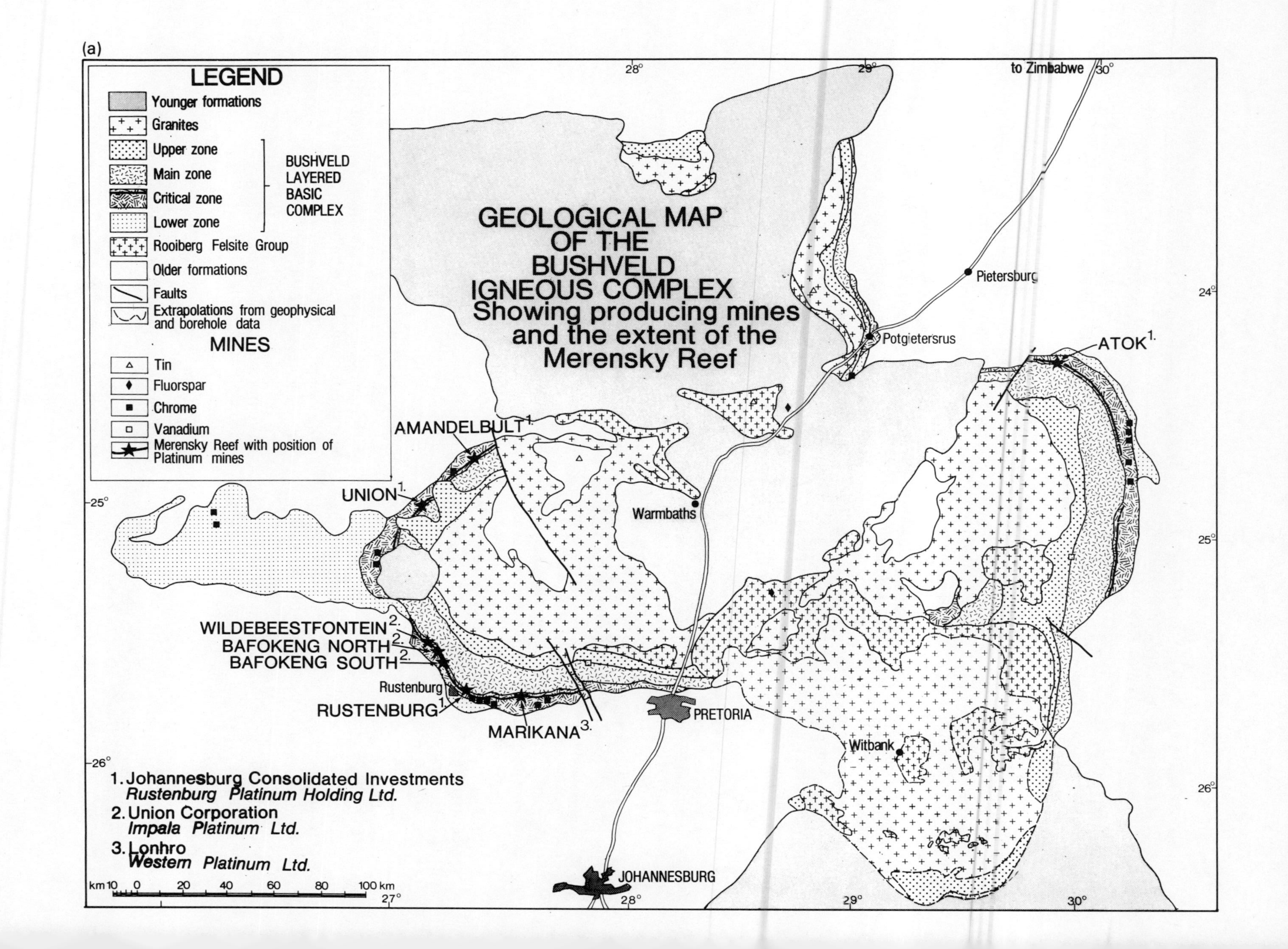

LEGEND
Younger formations
Granites
Upper zone
Main zone
Critical zone
Lower zone
BUSHVELD LAYERED BASIC COMPLEX
Rooiberg Felsite Group
Older formations
Faults
Extrapolations from geophysical and borehole data
MINES
Tin
Fluorspar
Chrome
Vanadium
Merensky Reef with position of Platinum mines
GEOLOGICAL MAP
OF THE
BUSHVELD
IGNEOUS COMPLEX
Showing producing mines
and the extent of the
Merensky Reef
to Zimbabwe
Pietersburg
Potgietersrus
ATOK 1.
AMANDELBULT 1.
UNION 1.
Warmbaths
WILDEBEESTFONTEIN 2.
BAFOKENG NORTH 2.
BAFOKENG SOUTH 2.
Rustenburg
RUSTENBURG 1.
MARIKANA 3.
PRETORIA
Witbank
JOHANNESBURG
28°
29°
30°
24°
25°
26°
27°
1. Johannesburg Consolidated Investments
Rustenburg Platinum Holding Ltd.
2. Union Corporation
Impala Platinum Ltd.
3. Lonhro
Western Platinum Ltd.
km 10 0 20 40 60 80 100 km

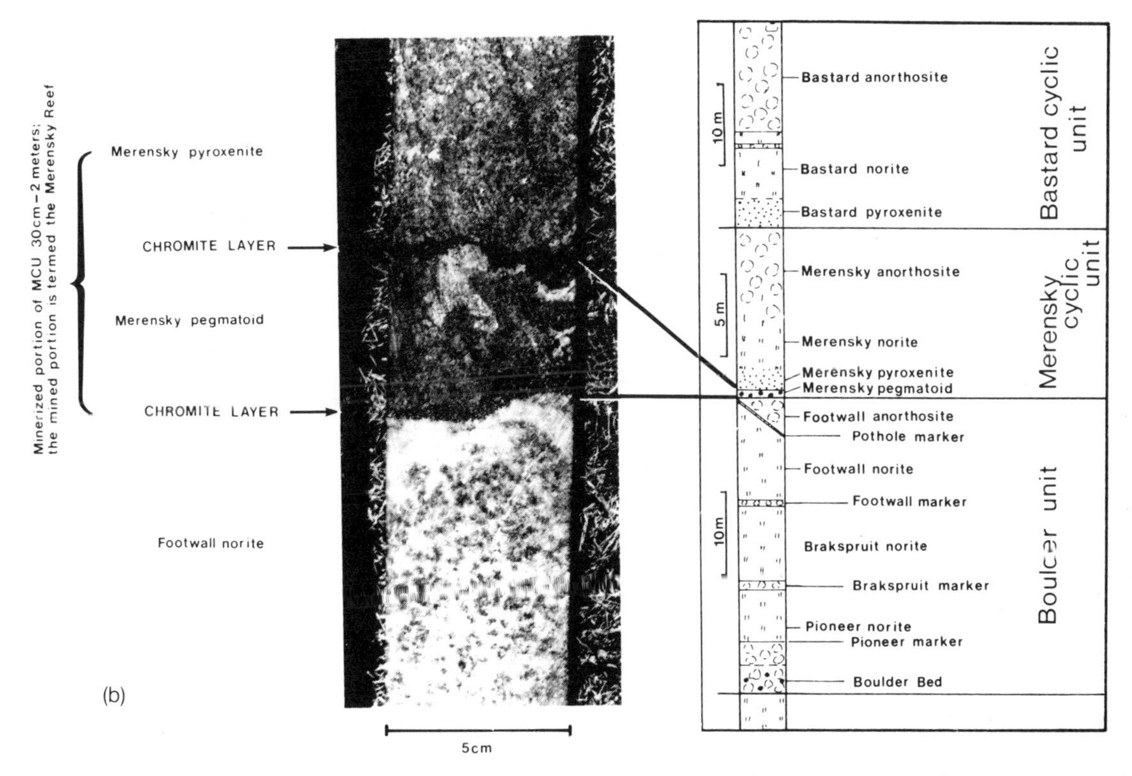

Fig. 2.5. (a) Simplified geologic map of the Bushveld Igneous Complex, with its producing mines. Also shown is the extent of the Platinum Group Element enriched Merensky Reef. The eight operative platinum mines shown, are controlled through three mining companies (see lower left-hand corner). Companies 1 and 2 produce the bulk of the platinum output (56 per cent and 41 per cent of the 2.14 million troy oz in 1979, respectively; Buchanan 1979*b*), and control a duopoly.

(b) Rock sample from a drill core (left; author) through the Merensky Reef at Rustenburg, compared to its local detailed stratigraphic setting (right; from Kruger and Marsh 1985). The term '**reef**' is not used for a specific layer of rock or rock type in the succession, but is reserved for whatever part of the Merensky cyclic unit (MCU) is mineralized or mined. The term **Merensky Reef** can therefore refer to the Merensky pegmatoid or the Merensky pyroxenite, or part of each. The reason for this usage is that in the literature and in the mining community, the **Merensky Reef** is that which is mined. The Merensky pegmatoid, which may vary in thickness from 0 to 2m, is often regarded as the reef, but may not be mineralized and indeed may be absent from the MCU, in which case the Merensky pyroxenite is the mineralised layer and therefore the reef (Kruger and Marsh 1985). For technical reasons, a minimum of 75 cm is mined.

petrological, and chemical data of its exposed sequences with their counterparts from the Bushveld Complex and then by further extrapolation of the Bushveld data to the 'hidden parts' of the Dufek complex.

As can be seen in Fig. 2.6, the Bushveld has been subdivided into five zones; the Basal and Transitional Zones which are chromite-bearing pyroxenites, dunites and harzburgites, followed by the Critical Zone which consists predominantly of pyroxenite (bronzitite), norite and anorthosite together with over 20 well-developed chromitite layers. The important platinum-bearing Merensky Reef is present at the top of this Zone. The Main Zone is mostly gabbro and norite while the Upper Zone consists of magnetite-bearing diorites and anorthosites.

The Dufek Massif section of the Dufek intrusion is mainly gabbroic in character while the Forrestal Range is dioritic with abundant magnetite. Clearly these two sections can be broadly correlated with the Main and Upper Zones of the Bushveld, respectively.

The chemistry of the pyroxenes in the Dufek rocks has been outlined by Himmelberg and Ford (1976) while data from the Bushveld are given by Atkins (1969) and Buchanan (1978). The chemical trends are very nearly the same for the Bushveld and the Dufek massif (Fig. 2.8).

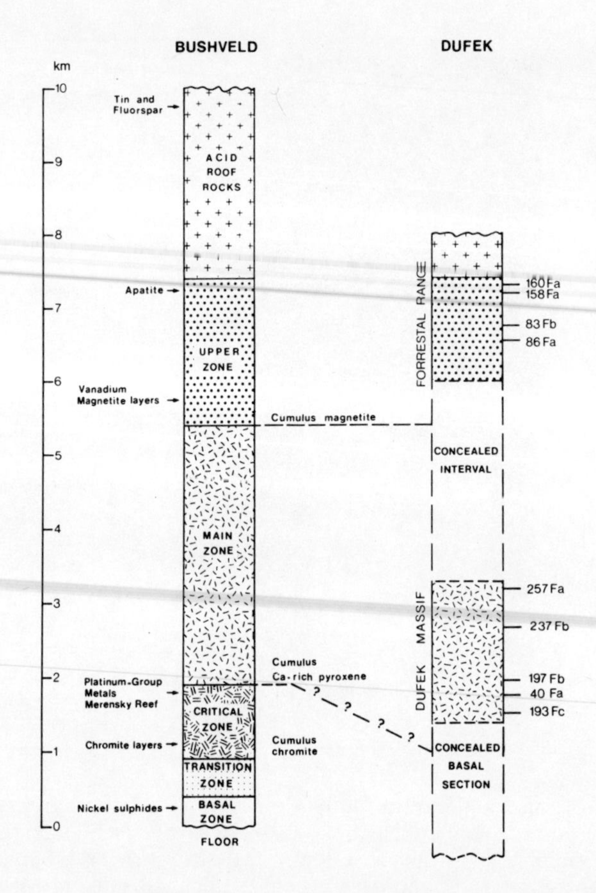

Fig. 2.6. Stratigraphic comparison of the Bushveld (South Africa) and Dufek igneous complexes. Mineralization in the Bushveld complex occurs at the indicated stratigraphic levels. The numbers on the right of the Dufek column are the rock samples (taken from Himmelberg and Ford 1976) for which coexisting pyroxene compositions are plotted in Fig. 2.8. Both these diagrams illustrate that if platinum mineralization comparable to that of the Bushveld complex (e.g. the Merensky Reef, the Platreef, the UG_2 zone) is to be found within the Dufek complex, it would most probably occur close to the top of the concealed Basic Section. The Dufek platinum metals mine model of this study is based on the assumption that a Merensky-type reef is present in the upper 1000 m of this concealed section.

The variation of the chemistry of the Bushveld pyroxene pairs with stratigraphic height is also shown in Fig. 2.8 relative to the compositional field of the platinum-bearing Merensky Reef. As can be seen the fractionation sequence in the lower portion of the stratigraphic column is contracted compared to the Upper Zone rocks. There is, however, a very good compositional spread where the Bushveld rocks overlap with the Dufek samples (nine of these have been plotted on Fig. 2.8; their locations are indicated in Fig. 2.6). They confirm that the Dufek section is broadly comparable to the Main Zone of the Bushveld with the Forrestal section showing iron enrichment trends similar to those of the Upper Zone.

Sample 193Fc was taken from the lowest exposed part of the Dufek section, just before the rocks disappear below the snow and ice cover. It is clear from Fig. 2.8 that it remains above the equivalent position of the Merensky Reef. It is reasonable to postulate, therefore, that rocks equivalent to those present in the Merensky Reef exist within the concealed basal section at no great depth, and since the layering of the complex is slightly tilted upwards to the west, such a reef may occur close to the surface somewhere to the west of Cairn Ridge and Hannah Peak (Fig. 2.4). This conclusion is used in the following section as a basis on which to evaluate the technical and economic feasibility of an Antarctic platinum mine in this region. For

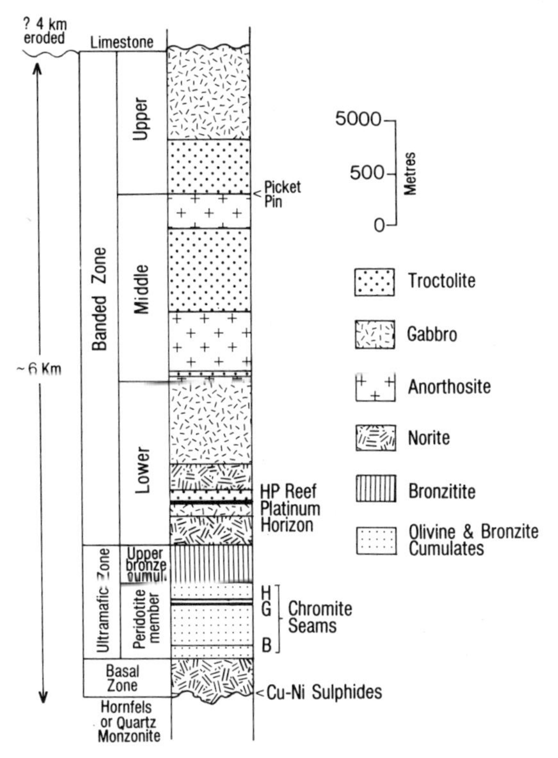

Fig. 2.7. Generalized stratigraphic section of the Stillwater complex, showing the location of the HP platinum enriched reef and other known mineralization (after Campbell 1984), for comparison with the Bushveld and Dufek sections.

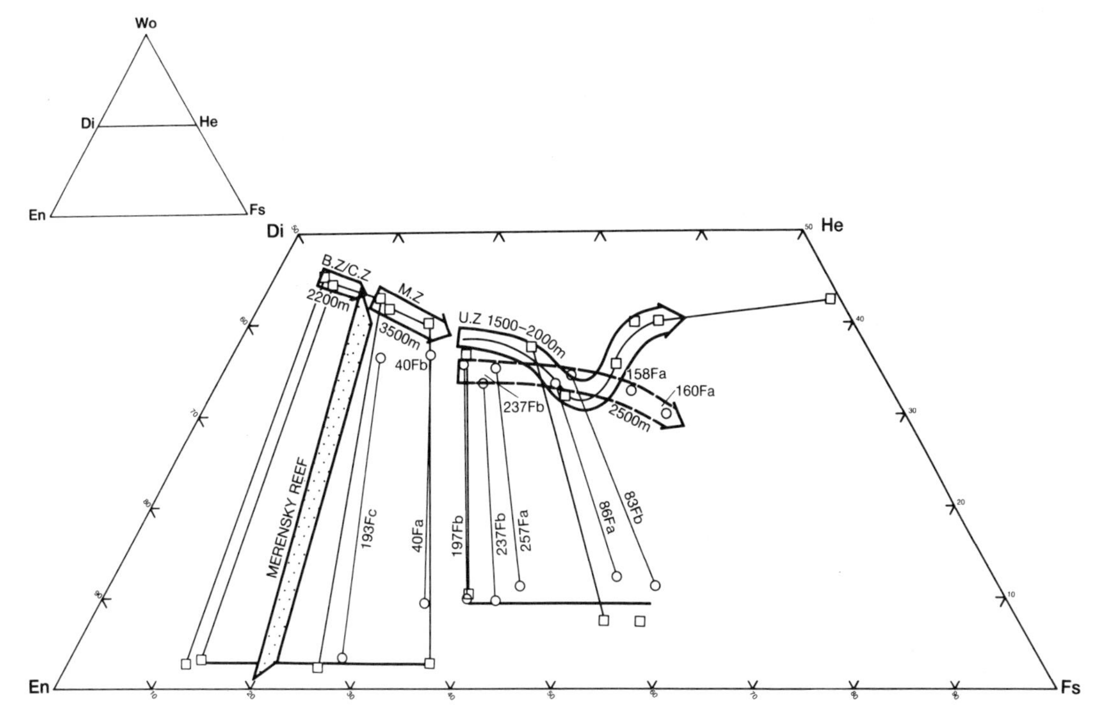

Fig. 2.8. Comparison of the compositions of coexisting pyroxene minerals from the Bushveld (squares) and Dufek (circles) igneous complexes. The broad arrows illustrate the gradual chemical changes of these pyroxenes from the bottom to the top of these complexes. The important platinum-bearing Merensky Reef of the Bushveld complex, lies below the lowest recorded pyroxene samples from the Dufek complex. Dufek rocks which should contain pyroxenes similar in composition as recorded from the Merensky Reef, are hidden beneath the Antarctic ice.

simplicity, small quantities of nickel, copper, and gold which are also expected to be present will be regarded as sweeteners. In this study, however, chromium will not be considered, even though chromite horizons may also contain additional favourable platinum contents which, through recently developed technology, can be easily separated (Corrans *et al.* 1982). For example, the UG-2 chromitite reef of the Bushveld complex which contains nearly double the amount of platinoids compared to the exploited Merensky Reef, is at present being developed for exploitation.

The important geological facts for this study are that the concentration of PGM in the Bushveld complex occurs in a thin (<1 m) and well defined layer. This Merensky layer can be traced for hundreds of kilometres along strikes in different parts (possibly different irruptive units) of the complex (Fig. 2.5). Similarly, it is now known that ore-grade PGM is also concentrated in such a thin layer in the Stillwater complex. This H.P. layer occurs at a comparable stratigraphic level to that of the Merensky Reef (Fig. 2.9). Additionally, in both complexes PGM is held in sulphides. Indeed there is such a striking similarity between the type as well as the stratigraphic positions of the PGM mineralization in these two complexes, that several generalizations *vis à vis* the level at which such PGM mineralization can be expected to occur, have come to light. Both the Merensky and the H.P. reefs, for example, occur about 500 m above the first appearance of cumulus plagioclase. A recent model for platinum-rich sulphide horizons postulates that such deposits must occur between 400 and 1000 m above the level of this first appearance of cumulus plagioclase in a layered complex (Campbell *et al.* 1983). The fact, therefore, that plagioclase is still present as a cumulus phase at the bottom of the Dufek complex before it disappears below the Antarctic ice, is consistent with the pyroxene evidence that a PGM sulphide layer should occur below this level. Assuming an average thickness of the hidden zone of 2.6 km, and a reasonable thickness of 1.5 km of ultramafic rocks at the base of the Dufek complex, then a Merensky-like reef can be expected to occur between 100 and 700 m beneath the lowest-known exposure level (Fig. 2.9).

Further evaluation of the potential PGM mineralization within the Dufek complex, as well as the expected platinoid-ore grades, will be discussed following the analysis of a hypothetical Antarctic platinoid mine (hereinafter

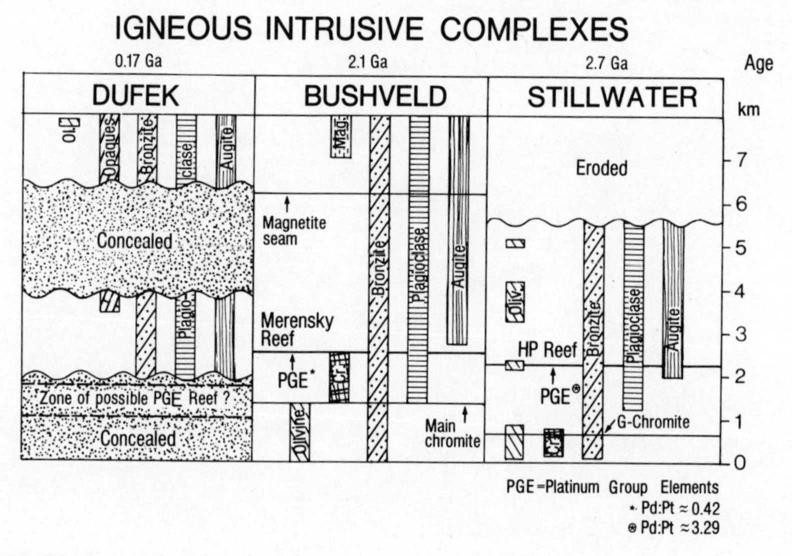

Fig. 2.9. Comparative average stratigraphic columns of the Stillwater (2.7 Ga), the Bushveld (2.1 Ga) and the Dufek (0.17 Ga) complexes, drawn to the same scale, showing the vertical extent of the major cumulus phases, as well as the stratigraphic positions of the platinum group element enriched reefs of the first two complexes (data from Campbell, Naldrett, and Barnes 1983; Ford 1983) and that of the postulated position of such a reef, as argued in this study, within the Dufek complex. Note also that the higher Pd:Pt ratio in the older Stillwater complex as compared to that of the Bushveld complex may counter arguments that this ratio in the earth's mantle may have increased with geological time.

referred to as the Dufek Platinum Metals mine), inferred from the given geologic data.

2.2.3. Evaluating the economics of developing The Dufek Platinum Metals mine

General setting of the mine. The mine will be situated along the north-west face of the Dufek mountains, in the vicinity of Cairn Ridge and Hannah Peak (82°34′S 53°W; Figs 2.4 and 2.10), where the terrain, at an elevation of about 600 m above sea level, is relatively flat, ice-free and easily accessible from the north. A hard runway to accommodate large cargo aircraft (Hercules L-100-30) as well as jet airliners, can be constructed on the ice nearby. Scree slopes of the Dufek ranges offer sufficient loose rock for building foundations, and construction will be entirely on solid ground (Fig. 2.1)

The terrain grades gently (1 per cent) from the mountain face of the Dufek ranges 50 km northwards to the Ronne and Filchner Ice Shelf. From there, the shelf extends some 400 km at nearly constant elevation to the ice-front of the Weddell Sea. Ice thickness of the shelf ranges from less than 250 m to a little over 1000 m (Fig. 2.10). Careful observations of satellite imagery and air photographs indicate that relative crevasse-free routes can be established either to the west or to the east of Berkner Island over almost level ice (Robin, Doake, Kohnen, Crabtree, Jorden, and Moller 1983; Fig. 2.10). A permanent 550 km ice road will be constructed along the eastern side of Berkner Island, linking the mine site to the ice-shelf front. The costs of building and maintaining such a road on a year-round basis, including the building of possible ice-bridges, have been estimated using data available from the construction and maintenance of winter roads on large lakes in Northern Canada (Adam 1979). The edge of the ice-shelf may be up to 60 m in height (e.g. north of Berkner Island), but is generally much less (Robin *et al.* 1983). Docking and loading facilities will be established through exploitation of natural features at the ice-front. For instance ice ramps, formed within extension fractures at right angles to the ice-front, will be reshaped and tailored to suit the needs of this project, as is also done during unloading of equipment for scientific expeditions and the building and refurnishing of scientific stations along the ice-front (R. Adie, personal communication 1983). Loading and storage facilities will need to be flexible on a long term basis to accommodate periodic ice-calving at the front of the shelf.

Weather conditions at the ice-front are favourable for shiploading periods from 8 to 10 weeks during the southern summer months (December to February). Argentinian and Soviet ships service Belgrano and Druzhnaya stations, respectively, during this period. Recently a new West German summer base along the Weddell Ice Shelf (Fig. 2.10) was established, using the Norwegian ship 'Polar Queen', during a period from December 1981 to January 1982. Nevertheless, to avoid the risk of failing to reach the intended docking–loading base, at least one concentrate-cargo ship of ice breaker class will have to be available.

Meteorological observations over periods of more than two decades have been collected at the nearby scientific stations Belgrano and Halley Bay, and synoptic observations were made during the Trans-Antarctic Expedition (1955–1958) at Shackleton Base and South Ice Station for 1.5 and 0.8 years, respectively. Some of these data are summarized in Table 2.1 and 2.2. Mean air temperatures over the three summer months range from −3.0 to −3.6°C, with maximums of up to +3.4°C and minimums as low as −29.6°C (measured over a period of 27 years at Halley Bay; British Antarctic Survey data, private communications with D. Limbert 1983). Wind speeds average 19 to 12 knots, whilst the probability of gales during these months is extremely low (0.03). Precipitation at Belgrano is about 30 cm per year (4 cm in January, Table 2.2).

Weather conditions at the mine site are more akin to those of the coastal regions than to the harsher climate of the interior high plateau (2000 m), which includes most of Antarctica (Lister 1960). Additionally, the mountain ranges partially shelter the mine site from the interior catabatic (gravity-driven) winds. Precipitation is low and mean temperatures are not unlike those at the Polaris mine in Canada, whilst winter lows are similar to those experienced at the Lupin Gold mine, Northwest Territories, and in the oil districts of Northern Alberta, Canada. For example at South Ice Station (81°57′S 28°52′W) the lowest temperature recorded was −57°C on 29 July 1957, and the highest was −8.2°C on 13 December 1957. Accumulation of snow, as measured in pits and from cores was approximately 10 g/cm^2 at the turn of the century and reached a peak of 17 g/cm^2 in about 1920, but

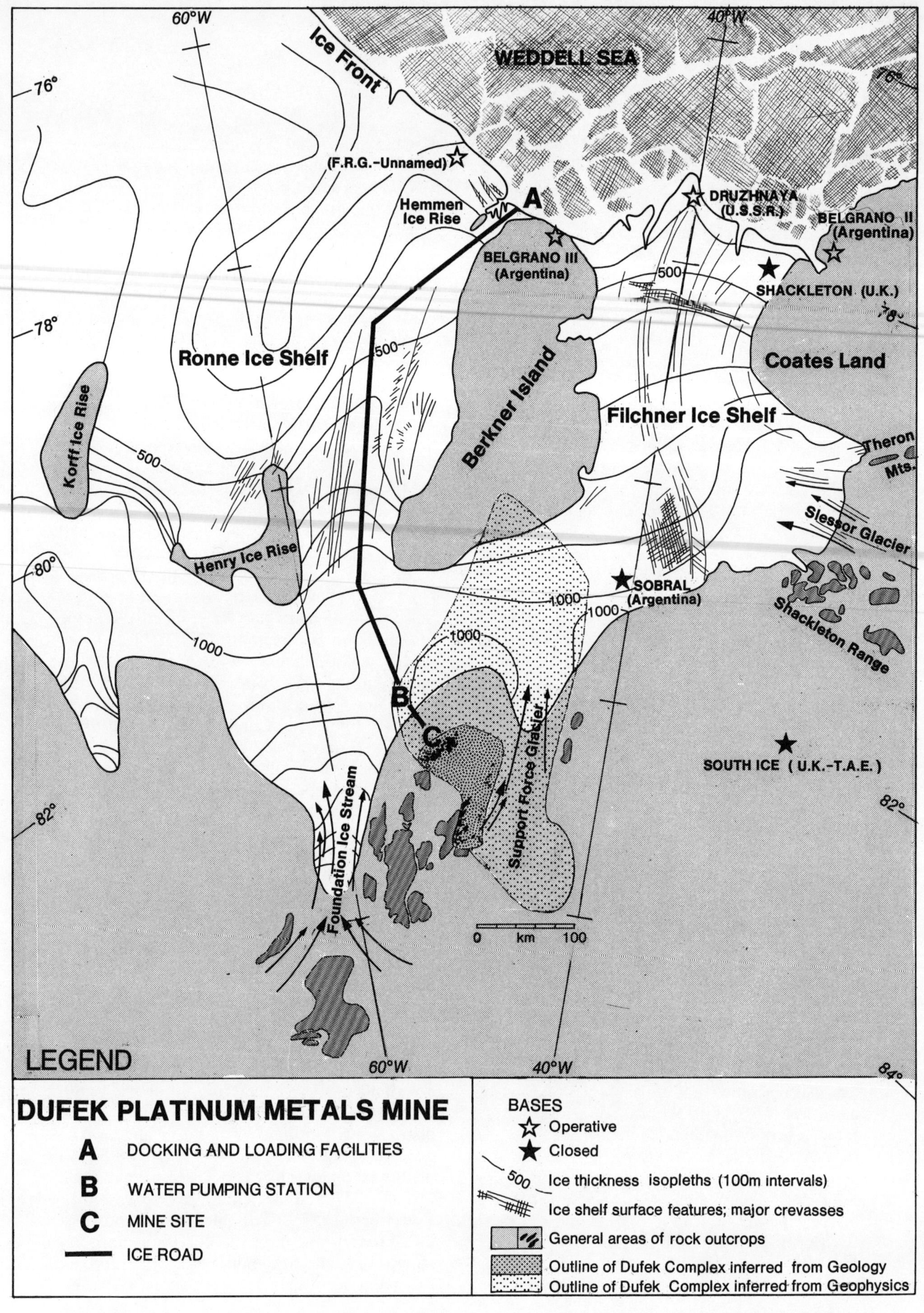

Fig. 2.10. Location map of the Dufek Platinum Metals Mine, Antarctica. The ice road which links the ice front of the Ronne Ice shelf to the proposed mine site, covers about 550 km. The road avoids the major known crevassed sections of the shelf. Docking facilities along the Weddell Sea border (A) allow for ship-loading periods of up to 8–10 weeks per year (December–February). A deep submersible water pump installed at B will supply the mine with sufficient water for plant operation and domestic use. Ice thickness isopleths taken from Robin, Doake, Kohnen, Crabree, Jordan, and Moller (1983).

TABLE 2.1. *Monthly temperature summary (°C)—Halley Bay Station, UK*

Year/Month	Extreme		
1981	Mean	Max	Min
January	−3.7	+0.1	−12.6
February	−8.6	+1.2	−20.0
March	−16.2	−1.6	−41.3
April	−17.2	−0.4	−38.5
May	−26.9	−11.3	−43.7
June	−25.9	−10.8	−42.3
July	−24.7	−11.5	−42.9
August	−25.1	−8.6	−40.7
September	−26.2	−12.1	−45.7
October	−18.7	−9.2	−28.0
November	−10.6	−2.2	−19.7
December	−4.8	+2.0	−12.9
1982			
January	[illegible]	+3.3	−16.5
February	−8.0	+0.7	−18.6
March	−21.7	−8.1	−35.4
April	−18.9	−3.9	−33.8
May	−25.4	−9.4	−45.0
June	−26.4	−7.9	−46.8
July	−24.7	−8.3	−42.6
August	−28.7	−10.6	−48.7
September	−20.3	−10.0	−38.7

Mean of reports at 00, 03, 06, 09, 12, 15, 18, and 2100 h GMT. Source: British Antarctic Survey (1983).

TABLE 2.2. *General climatological summary—Belgrano base (Argentina)*

Temperature	
Mean annual (period)	−22.2°C (13 years)
Highest mean monthly (month)	−6.0°C (Jan–Dec)
Lowest mean monthly (month)	−32.8°C (August)
Absolute maximum (date)	9.2°C (28 Dec. 1975)
Absolute minimum (date)	−59.5°C (July 1972)
Wind	
Mean annual velocity (period)	9.3 knots (13 years)
Cloudiness	
Mean annual (period)	$4\frac{7}{8}$ (13 years)
Highest mean monthly (month)	$5\frac{5}{8}$ (October)
Lowest mean monthly	$3\frac{5}{8}$ (June)
Overcast days	46%
Part overcast days	33%
Clear sky days	21%
Precipitation (equivalent in water)	
Mean annual (period)	30.8 cm (6 years)
Highest mean monthly (month)	4.14 cm (January)
Lowest mean monthly (month)	1.2 cm (July)

Data covers period 1955–1979. This station was closed down on 30 January 1980 and is being gradually replaced by stations Belgrano II and Belgrano III. Source: British Antarctic Survey (1983).

declined to 10 g/cm^2 in the 1940s, dropping further to 6 g/cm^2 by 1957. This gives a half century average of 12 g/cm^2/year (Lister 1960).

Exploration and ore reserves. In practice, the difficulty in defining the end of an exploration programme and the start of exploitation bears significantly on technical and economic mine-modelling. Two years of exploration are therefore allowed for prior-to-project approval to develop the hypothetical mine. Arguably, with the present state of geologic knowledge about the Dufek massif such a period appears justified, although normally under conditions of high risk, exploration cannot be economically compressed into such a short time frame.

The bottom line of the available geological data indicates that only with significant amounts of drilling will any further advances be made towards realistically evaluating the Dufek Complex as a potential mining opportunity. Hard-rock drilling of up to a total length of 20 000 m is therefore planned. Logistics for such an exploration period will be supported by air from southern South America using Hercules ski-equipped aircraft (144 round trips). Flying time needed and costs to maintain base-camp facilities have been calculated from data gathered during building of large semi-permanent camps for geological expeditions in the nearby Ellsworth Mountains (Fig. 1.1, Splettstoesser, Webers, and Waldrip 1982; Splettstoesser, personal communication 1982; G. Guthridge of the NSF Office of Polar Programs, personal communication 1983). Expenses for salt, needed to depress freezing temperatures of drilling fluids, were provided by Cominco from their drilling experiences through the Greenland ice cap (private communications 1982, 1983). We will assume that this exploration programme guarantees the economic potential of the deposit. Thus, the economic evaluation of the Dufek opportunity as it exists now does not take into account the risk that nothing of potential economic value will be found.

The ore horizon can be assumed to lie close to the surface in the proposed mine area, as projected from the mineralogical data and the geological field measurements of the dips of the layers of the complex in the Dufek area, which range between 0° and 8° to the south-east (Fig. 2.4). The shallow dip of the rock ensures that underground mining will be possible for at least 20 km into the complex, over a strike length of more than 15 km and probably up to 50 km. This implies vast ore reserves, easily accessible to a simple mining method. Typical ore grades are given in Table 2.3. They are similar to those found in the Bushveld complex (compare with Tables 2.7 and 2.11).

Technical aspects. Development of an underground mine using an open-stoping room and pillar method appears to be the most feasible and practical mining approach, with the advantage of simplicity of operation and ease of mechanization. The extent of underground mechanization will be dependent on the thickness of the ore horizon, which is not known, but is here assumed to be of the order of 3 m. This will allow a highly mechanized operation. Initial mine production will be modelled to accommodate such a thin ore horizon, similar to that at Polaris, which itself is based on methods in use at the Missouri lead belt mines. Ore mined will be of the order of 750 000 tonnes per year at full production. A production period of 26 years was set in order for the mine to process some 15 million tonnes of ore reserves (see Table 2.3 for further details). No mine recovery factor is specified; it is assumed that there are 'infinite' reserves availabe in the Dufek complex considering the relatively small size of the mine.

TABLE 2.3. *Dufek Platinum Metals mine specifications*

Depth to ore body (m)	<1000
Dip of ore body	<10°
Ore width (m)	<3
Ore reserves (million tonnes)	>15
Typical ore grade (PGM and Au) (g/tonne)	8.1
Platinum grade (g/tonne)	4.82
Mine life (years)	26
Stope width (m)	3
Ore dilution (%)	22.3
Typical head grade Pt (g/tonne)	3.94
Ore milled (tonnes/day)	2050
Concentrate produced (tonnes/day)	331
Mill recovery (%)	78
Refining recovery (%)	98
Operating schedule (days/year; 8-h day)	
Underground	250
Concentration plant, general facilities	350
Manpower requirements	
(No. of employees, total)	225
Mine	77
Mill	33
Surface	63
Administration	52

A partial incline–decline will probably be available from the two-year period of exploration, and pre-production mine development will involve its enlargement. A shaft may not have to be built, but for conservative measures the costs for construction of such a concrete shaft, about 5 m in diameter to a depth of 1000 m, has been allowed for in the economic feasibility study. It is proposed to install an underground crusher and conveyor system to haul the ore to a stockpile on the surface. Raises will be bored from the surface to provide the mine ventilation system. Underground services will include a garage with equipment servicing facilities, a warehouse, rest and first-aid rooms, electrical sub stations, explosive storage facilities, etc. To keep underground labour to a minimum, mining equipment will be of high productivity; machines will be rubber-tyred, of high capacity and feature heated operator cabs.

The process facilities will have the capability to handle 2050 tonnes per day. (Table 2.3 and 2.4). Mill capacity will reach 100 per cent over a period of 4 years. A prefabricated process plant with power and service facilities will be installed at the mine site, following towing to the ice-shelf from Punta Arenas, southern Chile, and haulage across the ice road on a sledge-like bottom surface (Fig. 2.10). Design allowance will be made to facilitate the addition of new sections to the process facilities, should it become desirable to increase capacity. A process equipment flowsheet for complex base metal ores is assumed. The mill will be using salt water in the flotation circuits.

The concept of building the facilities on barges in southern South America (Chile and/or Argentina, both of which have good port and air facilities) and having them towed to the site, has evolved from the Polaris experience as a means of minimizing on-site construction. In addition, equipment to be installed on-site will be prefabricated and shipped or flown in maximum sections. For example, the conveyor systems for the crushing plant, concentrate storage, and loading systems will be sectionalized so that local work will only consist of component assembly. Sea and air transportation distances from Punta Arenas to the mine site (*c.* 4200 km and *c.* 3300 km, respectively) are comparable to those between Montreal and Polaris, whilst concentrate transportation from the ice shelf to a smelter would be over a comparable distance to that over which concentrate is transported

TABLE 2.4. *Dufek Platinum Metals mine: project flow sheet*

Year	*Month*	*Significant features of schedule*
1–2 (1984–85)	Nov.–Jan.	Initial camp and airbase facilities prepared. Exploratory drilling completed (20 000 m).
3 (1986)	Nov.–Jan.	Project approval; initial airlift of construction equipment, supplies, and personnel (100). Existing exploration camp refurbished. Roads, docking, surface mine area, and accommodation sites prepared prior to the shipping season of the year.
4 (1987)	January	First ocean shipment of construction equipment and materials to allow for major start of development work. Additional construction camp will be brought in at this time to supply accommodation until permanent housing is ready for occupation in year 4.
4	July	Opening of existing underground workings form exploration period, underground development and most of the surface work done.
5 (1988)	January	Concentrator complex completed in southern hemisphere (S. America; Japan; South Korea) in mid-year 3, moved in. Mine development work completed and undergound crusher and conveyor system installed. Mine production starts as development is phased out and ore delivered to stockpile at plant site.
5	July	Start up of concentrator and delivery of concentrate to stockpile ready for first shipment.
6 (1989)	January	First shipment of concentrate produced with mill capacity at about 70 per cent.
7 (1990)	January	First shipment of concentrate produced with mill capacity at about 80 per cent.
8 (1991)	January	First shipment of concentrate produced with mill capacity at about 95 per cent.
9 (1992)	January	First shipment with mill capacity at 100 per cent. reaching 2050 tonne/day.

between Polaris and Antwerp in Belgium. Transportation costs have been calculated accordingly (Table 2.5).

Oil will be stored in large rubber bladders as is common at the Williams aerofield, McMurdo (*Antarctic Journal of the United States* 1983) and many other Antarctic stations. Accommodation (consisting of wooden structures to cater for 240 men and supply recreation, dining, and office facilities) and concentrate handling facilities will be constructed on site. During the early pre-production years, a portal camp will be established, refurnished from the existing exploration facilities (which include a provisional airstrip), to accommodate 100 people for the initial construction period.

From the Arctic experience, one of the key aspects of this operation is the ability to attract and hold a stable work force; good accommodation is therefore an essential consideration. Novel, prefabricated, cost-efficient accommodation designed for the extreme conditions in the Antarctic is now readily available. For example, despite their very low budget allowance, the British Antarctic Survey have recently been able to install a flexible tube-like building of wood able to withstand even harsher conditions than at the Dufek site (Relph-Knight 1982; B.A.S. Annual Report 1981–1982 and personal communications 1983). It is assumed that the labour force will be drawn from southern Chile and/or Argentina, whereas professional expertise will initially be hired in the northern hemisphere.

Concentrate will be carried in trucks across the ice-road to wooden domes with eleven months' storage capacity at the docking site near Berkner Island, and a concentrate loading system of 1500 tonnes/hour will be installed. Concentrate will then be shipped to smelter and refining facilities in Brazil, Australia, Japan or the southern United States.

Water will be pumped to the mine site over a distance of about 50 km, using galvanized steel pipes, from a suitable site on the ice-shelf (Fig. 2.10). This may involve drilling through up to 1000 m of shelf-ice to tap the seawater below, but enough experience has been gained by now, in Greenland as well as Antarctica, to drill through ice and permafrost conditions. (Longyear, private communications 1983; in 1975 Longyear supervised the NSF ice-drilling programme in Antarctica). A high quality, heavy-duty submersible mining pump of the type developed by RITZ Pumpen Fabrik, West Germany, for use in deep mining in South Africa and elsewhere, will be installed. RITZ is confident that such a pump, with a head of between 800 and 1000 m, can be tailor-made at a cost of between US $ 0.2 and 0.5 million and delivered in time for the specific conditions set in this study (private communications 1983). Water at the mine site will only be desalinated for domestic consumption and limited plant needs, using the diesel engines of the mines' power system. Mill tailings and sewage will be pumped into the Weddell Sea water below the ice-shelf. A water recycling system operating within the mill will increase the specific gravity of solution discharging considerably; this will reduce the 'fines plume' effect caused by the rising of less dense water in tailings discharge. The mine will operate using dry drilling techniques in the underground ore stoping, similar to those at Polaris. This will eliminate the production of soluble heavy metals. When underground work reaches operational status, it is expected that empty stopes will be filled with the tailings and allowed to freeze. This should minimize rock-mechanical problems below surface and abate the risk of environmental hazards above ground. Ecological baseline studies will be carried out from the start of the project, 6 years prior to production.

Project and production schedule. Critical constraints for a successful mining operation are the short Antarctic shipping seasons, and the decision-making ability to meet a tightly knit schedule within the confines of such an eight- to ten-week access period. For example, a delay in the building of the process plant will result in a minimum of one year's delay in concentrate delivery. The annual quantity of concentrate to be shipped can be handled on a competitive basis by Canadian, Swedish, Finnish or Belgian ice-breaking and ice-stengthened cargo vessels, since they should be available at the critical shipping time during the austral summer when activities in the Arctic are at a minimum.

The proposed mine development can be completed within 4 years from the time of decision to proceed, and full production can be planned for within a further 3 years. Key dates are as follows:

Year 1	November/ January	Project approval, initial airlift and start of field work
Year 2	January	Sealift of all on-site material
Year 2	July	Start mine development and construction plant
Year 3		Peak mine development; concentration plant built
Year 4	January	Move process plant to mine site
Year 4	July	Start up concentrator
Year 5	January	First shipment of concentrate from mill capacity at 70 per cent; underground production 74 per cent
Year 6	January	Mill capacity at 80 per cent; underground production 82 per cent
Year 7	January	Mill capacity at 95 per cent; underground production 96 per cent
Year 8	January	Mill capacity at 100 per cent (2050 tons/day); underground production 100 per cent (2392 tons/day).

A more comprehensive scheme is given in Table 2.4.

Economic concepts, models, and scenarios. Given the geological constraints and assumptions, several economic models simulating the cash flow throughout the life of the Dufek Platinum Metals mine have been constructed, enabling a broad assessment to be made on the potential profitability of this mine. Modelling was conducted with two fundamental questions in mind. *Firstly,* what order of profitability can be expected given a reasonable set of invariant geological constraints, including ore-grade and composition, and how would such profitabilities change with variable geological and economic parameters? *Secondly*, what would be the sort of ore-grade needed to contemplate establishing a viable mine in the vicinity of the Dufek complex? It is only with such knowledge that further meaningful discussions about the likelihood of finding profitable grades can be conducted.

The profitability of a project is the measure of its worth to an investor over the total life of the project. There are a number of ways of evaluating profitability. Most are related to the concept of cash flow. The cash flow of a project is usually calculated annually, and is derived by taking the total revenue and subtracting all expenditures made during that year, as well as taxation and interest payments. Allowances, such as depreciation of fixed assets or depletion of exhaustible resources, are subtracted for the tax calculation, but are added back to give the resultant cash flow. Discounted cash flow analysis (DCF) is a common project evaluation technique. It is essentially the reverse process of the more familiar concept of compounding used to calculate future savings. Predicted cash flows over the life of a project are re-calculated to their present value by applying a suitable discounting rate. Discount rates used in project analysis take into account not only a time value of money (money earned in the future is worth less in present day terms) but also the inherent risk factor of the project. The discount rate of a project may be referred to as the cost of capital and can be easily weighed up against the use of that capital in other investment opportunities. Mining is a high risk venture and consequently analysts of mining projects tend to use high discount rates in their feasibility studies. An initial discount rate is usually set equal to a desired minimum rate of return.

The overall real returns to mining companies are probably of the order of 5 per cent. This figure will take into account all ventures, successful and otherwise, together with exploration ventures. When considering an individual project, a discount rate of 8–10 per cent is more appropriate, since the cash flows will only pertain to the mining project and a lot of the exploration expenditure will be sunk costs. Allowing for a 10 per cent annual rate of inflation, a real 10 per cent discount rate was chosen as a standard for the Dufek Platinum Metals mine project in order to incorporate some aspects of risk related to mining in such a remote and untested terrain.

There are three basis measures to evaluate the profitability of a project using DCF:

(a) net present value (NPV);
(b) internal rate of return (IRR); and
(c) Payback period.

Ideally, these measures should be assessed in concert for overall evaluation purposes, since they each convey different aspects and kinematics of a project which may bear on the short or long-term objectives of an investor. The NPV of a project is the sum of the present values of all future annual cash flows over the life of the project. The IRR is the value of the discount rate at which the NPV of a project is zero. Thus, if the IRR is less than the required discount rate, the project will not generate an acceptable return. Conversely, if the IRR exceeds the required discount rate, an acceptable return on investment will be generated. The payback period is the time taken to repay the initial capital outlay. It is calculated as the time taken from production, when the cumulative cash flow changes from negative to positive.

The basic elements of revenue, capital cost, operating costs and expenditure schedules used for the Dufek platinum metals mine are given in Tables 2.5–2.7. All costs are quoted in 1983 US dollars. Costs have been derived predominantly from the 1979 Polaris feasibility study (Cominco 1979) and, where appropriate, tailored and re-calculated to suit the Dufek Platinum Metals mine. Some of the specific technical differences relevant to this Antarctic mine are the incorporation of:

(a) a shaft;
(b) a more complex processing plant;
(c) additional overland transportation of the processing plant;
(d) additional overland transportation of the concentrate;
(e) building and maintaining a more elaborate infrastructure;
(f) Water supply to the mine-site from below the ice shelf.

These extra facilities can easily be accommodated using 1983 technology and, as will be shown later, none of the above have a significant influence on the overall economics of the mine. The basic capital investment (CAPEX) for the

TABLE 2.5. *Dufek Platinum Metals Mine costs*

Capital expenditure (CAPEX)	1983 US $ millions
Mine development	
Shaft, hoist, headframe, equipment	11.5 [1,2]
Transportation to site: 30 per cent of shaft etc. CAPEX	3.5
Underground equipment and rock development	25.34
Extras and escalation: 20 per cent of above	
Concentrator plant and surface facilities	
General (clearing, foundations)	21.95
Crushing, crushed ore handling	4.54
Concentrator	
Site and building	8.075
Power plant	8.075
Processing Plant	8.075
Sledge	0.775
Transportation to ice-front (3000km)	0.825
Transportation to mine-site (550km)	0.125
Concentrate storage and loading facilities	10.575
Trucks ($350 000/50 tonne truck)	3.15 [3]
Extras and escalation: 25 per cent of above	
General	
Construction ice-road ($716/km), bridge, ramp	0.394 [4]
Pipelines for water, sewage, tailings ($115.9/km)	0.356 [5]
Town site accommodation, recreation facilities	14.988
Development services (engineering, legal, management, fees)	15.525
Extras and escalations: 20 per cent of above	
Project overhead costs: 6 per cent of mine development and 8 per cent of remainder capital expenditure	
Exploration	
Drilling (20 000 m)	3.00 [5]
Salt ($200/tonne; 0.04 tonne/m drilled)	.16 [5]
General transportation (1340 flight h; $4000/h)	5.36 [6]

Operational expenditures (OPEX)		1983 US $—per day
Power		6714.00
Mine	Labour	10493.00
	Supplies	12138.00
Mill	Labour	4373.00
	Supplies	6483.00
Surface	Labour	10045.00
	Supplies	2383.00
General administration		17707.00
Transportation costs/tonne concentrate : mine to ice-front		73.15 [3]
Transportation costs/tonne concentrate : ice-front to smelter		68.75
Smelting charges /tonne concentrate		76.88
WORKING CAPITAL : 33.3 per cent of operational costs		

Sources: Comminco (1979) and authors' estimations unless otherwise indicated;
[1] R. Spencer, personal communication 1983.
[2] O'Hara (1980).
[3] J. Stocks personal communication 1983.
[4] Adam (1979).
[5] Comminco, personal communication (1983).
[6] NSF, personal communication (1983).

Dufex Platinum Metals mine can thus be confidently summarized as follows:

	($ millions) Constant 1983 $	Current $
Exploration CAPEX	8.5	9.8
Shaft and equipment CAPEX	17.9	28.0
Underground mining CAPEX	30.4	47.9
Treatment plant CAPEX	73.9	115.6
General facilities CAPEX	50.9	79.6
Project overhead costs	12.3	19.2
Total CAPEX	193.9	300.1
Working capital	9.8	188.3
Total investment	$203.9	$488.4

Since inflation has become a major aspect in economic assessment of projects today, variables used in the economic models have been inflated at separate specific rates, assuming a constant 10 per cent inflation throughout the life of the mine. The effects of this on the total capital investment are significant, as shown in the second column.

Two further structures are built into the Dufek Platinum Metals mine scenario developed so far. The first of these incorporates the effects of taxation on the project's cash flow and simultaneously allows economic analysis to examine a simple financing agreement. The second increases the elasticity of the model with respect to different platinum price paths and ore quality.

All models are set up either with or without taxation. In the no-taxation scenarios, the project values are the overall values. The latter are calculated with discount rates of 10 per cent and 15 per cent. These no-tax scenarios exploit a joint venture agreement between a government or an international agency (partner 1) and a commercial exploiter (partner 2). In all these models, partner 2, the corporate, takes all cash flow, bar a royalty, until it obtains a real return of 10 per cent. Thereafter, the cash flow is split 50–50. The royalty due to partner 1 is 5 per cent of the profits.

In the taxation scenario, a World Bank-type sliding-scale taxation structure is implemented (*cf.* Palmer 1980; World Bank 1982*a*). The parameters are as follows:

1. A standard corporate tax rate of 35 per cent on all taxable income. One hundred per cent of any capital expenditure is tax-deductible in the year that it occurs.
2. An internal rate of return surtax rate of 50 per cent on a real corporate IRR of between 15 and 20 per cent; 70 per cent surtax on a real IRR of between 20 and 25 per cent, and 80 per cent on a real corporate IRR greater than 25 per cent.

To complete the scenarios, four different price paths for platinum have been constructed, into which two variants of ore quality, representative of a possible spectrum of platinum to palladium ratios, are incorporated. The platinum (Pt) price scenarios are as follows:

1. $475/oz, kept constant over the mine life.
2. $475/oz, with 2 per cent per annum real Pt price increase from the start of the project.
3. $475/oz, increasing to $950/oz 2 years after the start of the project (e.g. in 1986), followed by a 10 per cent per annum real decrease over the subsequent 6 years.
4. $475/oz, increasing to $1520/oz 7 years after the start of the project (e.g. 1991), followed by a 10 per cent annum real decrease over the subsequent 10 years.

TABLE 2.6. *Dufek Platinum Metals Mine capital expenditure schedule*

Pre-production years	1	2	3	4
Shaft	10	40	30	20
Mine development	2.4	28.9	60.1	8.6
Concentrator and surface facilities	2.7	38.6	48	10.7
General	2.7	38.6	48	10.7
Operating costs	70	81.6	96.4	100.0

Expressed as a percentage of total expenditures

TABLE 2.7. *Dufek Platinum Metals Mine revenue distribution*

PGM	Ore grade (g/tonne)	Price (troy oz)	Refining charges ($/g)
Platinum	4.82	475.00	0.37
Palladium	0.4–1.6*	130.00	0.37
Ruthenium	0.14*	26.00	1.02
Rhodium	0.05*	300.00	1.02
Iridium	0.017*	300.00	1.02
Osmium	0.014*	130.00	1.02
Gold	0.054*	400.00	0.08
Base metals	weight (%)	($/tonne)	($/tonne)
Copper	0.28	1852.00	331.00
Nickel	0.17	7055.00	331.00

*Expressed in ratios of platinum grade

The last two scenarios thus create sharp peaks in the platinum price path at specific times during the project, followed by relatively long-term decreases. The first of these increases occurs during the pre-production years whilst the second coincides with the start of full production. These times were chosen for two reasons. Firstly, to expose the different effects of the project 'coming on stream' with respect to an external, dynamic platinum market, and secondly, to accommodate realistic expectation in future world platinum prices, as deducted from its growing status as an industrial commodity and its near monopoly-controlled production and marketing. The latter is discussed at length in the next chapter.

In each of the above scenarios, the palladium to platinum ratio of the mined ore is set either at 0.42 or at 1.6. The first corresponds to an average ratio in ores from the Merensky Reef in the Bushveld Complex (von Gruenewaldt 1977; Buchanan 1979*a* and *b;* Naldrett 1981, Sharpe 1982). The approximate four-fold increase to 1.6 is thought to cover expected variations in this ratio adequately (Buchanan, personal communications 1983), although more than eight-fold increases are recorded from the HP reef in the Stillwater Complex (Buchanan 1979*b;* Glacken 1981; Todd, Schissel, and Irvine 1979). Such variations must have significant effects on the revenues of any platinoid mine.

In summary, 16 basic models were constructed, and their relevant economic and financial parameters computed using the MECON programme of R. Spencer (Royal School of Mines, University of London; Glacken 1981; de Wit, Spencer, and Buchanan 1984*b;* see Appendix A5 for an example). The results are shown in Table 2.8. Each model was subjected to sensitivity analysis by flexing the values of the matrix parameters from +50 per cent to −50 per cent. The resultant changes of the values of some of the most critical parameters are plotted against the IRR for each model, to illustrate their relative impact on the rates of change of the profitabilities of the model projects (Fig. 2.11; 1–16). Table 2.9 lists the sensitivities of the models to each parameter tested in a more qualitative manner.

Profitabilities, sensitivities, and expectations. As intuitively expected, the results confirm that the possible profitability of the hypothetical Dufek Platinum Metals mine is closely tied to the platinum price. Surprisingly, however, it is also evident that given today's market prices (price scenario A), this mine might be a viable proposition even though, as structured in these models, the returns are too small or unattractive for a financial investor to consider the project seriously. Nevertheless, given the constraints, platinum *can* be profitably mined and the results are arguably encouraging enough to entice an enterprising mining company to scrutinize the technical and economic structure of this Dufek Platinum Metals mine model more closely. There is no doubt room for considerable economization and greater technical innovation, and given Model 3, a much tighter capital and opera-

TABLE 2.8. *Dufek Platinum Metals Mine profitability values of economic models*

Economic model	Price Scenarios PGM	Pd/Pt ratio	Tax	Net present value $ millions			Real IRR	Payback period	Joint venture: Government share ($ millions)		Joint venture: Corporate share ($ millions)		Return on corporate share	World bank-type taxation tax revenue ($ millions)			NPV-tax revenue ratio		
				0%	5%	10%	(%)		10%	15%	10%	15%	(%)	0%	5%	10%	0%	5%	10%
1	A	0.42		221	9.33	−50.8	5.5	7.53	4.22	2.33	−55	−67.4	< .6						
2			ON	117	−24.8	−63.2	3.57							104	34.1	12.5	1.1	−0.7	−5
3		1.6		458	108	− 3.83	9.72	5.52	6.59	3.50	−10.4	−43.6	≅8						
4			ON	282	46.9	27.4	7.49							176	60.7	23.6	1.6	0.8	−1.2
5	B	0.42		708	182	20.3	11.38	5.63	9.92	4.28	11.0	−36.0	≅11						
6			ON	452	98.1	−10.6	9.18							256	83.7	31.0	1.8	1.2	−0.3
7		1.6		1090	330	87.9	15.35	4.34	41.6	15.9	46.2	−12.8	≅14						
8			ON	713	204	39.4	12.85							374	126	48.5	1.9	1.6	0.8
9	C	0.42		288	38.4	−36.2	6.87	6.74	4.95	2.64	−41.2	−59.6	< 7						
10			ON	164	− 3.02	−51.7	4.84							124	41.4	15.5	1.3	−0.1	−3.3
11		1.6		545	145	15.0	11.18	4.98	7.53	4.03	7.44	−33.5	≅11						
12			ON	343	74.3	13.0	8.83							202	70.9	28.0	1.7	1.1	−0.5
13	D	0.42		701	283	117	21.73	2.06	53.7	27.4	63.4	15.6	≅17						
14			ON	333	137	50.2	16.85							368	148	67.0	0.9	0.9	0.8
15		1.6		1080	461	213	28.95	1.55	90.2	47.0	123	52.4	≅19						
16			ON	390	181	82.0	21.63							687	280	131	0.6	0.6	0.6

Price scenarios for PGM
A. $475/ounce platinum—constant.
B. $475/ounce platinum + 2 per cent p.a. real increase.
C. $475/ounce platinum, increased to $950 2 years before start of production, followed by a 10 per cent p.a. decrease for 6 years.
D. $475/ounce platinum, increased to $1520 2 years after start of production, followed by a 10 per cent p.a. decrease for 10 years.

tional budget might make this an interesting, if risky venture. Without any form of taxation, however, it remains an unlikely project (e.g. compare to Model 4). Alternatively, one might question the wisdom of this platinum price scenario. Such a constant price path over the next three decades, before the closure of this proposed mine in 2016, is equally unlikely. A real price increase of 2–3 per cent is thought to be a more realistic scenario by platinum experts (Chapter 3). This would make Models 5, 7 and 8 interesting propositions and even Model 6, which like Model 8 includes taxation, is worth considering, especially with better technological planning and greater economic stringency.

Interestingly, the price scenario C yields much less promising ventures. This can be directly related to the unfavourable time of project production scheduling vis à vis the considerable change in platinum price. The latter is modelled to double two years *prior* to full production of the proposed mine. The slow decrease following this peak price significantly alters the cash flow of the mine in a negative manner. This example serves to illustrate the importance of project timing and scenario planning in mining ventures where capital return is the prime objective. Inversely, such platinum price-hikes following the start of production may considerably increase the attractiveness of the project as an investment prospect. This is the case with the models using the fourth platinum price path (price scenario D). These models (13–16) could all be extremely attractive, both to a government (or an international agency) and a mining company as a joint venture, or for a mining company operating under a reasonably 'stiff' taxation scheme, as indicated by the NPV/taxation ratios (table 2.8). The real income to a national or international agency could run into hundreds of millions of dollars. Such a price scenario is not unduly speculative as verified in the next chapter.

The sensitivity analysis shows the mine to be very dependent on the platinum ore-grades, the underground mine dilution of this ore, hence mining methods used and effective grade control, and the total capital and operational expenditures. These are all important aspects to weigh up when assessing the risk factor and viability of each particular model. For example, to make Model 2 an attractive investment, given the base-case ore-grade, one would need to be able to decrease the capital and operational expenditure by more than 50 per cent, whereas in Model 15 one could effectively increase these expenditures by more than 50 per cent and still retain a break-even value. Similarly, for Model 2, much higher ore grades need to be found relative to Model 15, whilst the latter model could withstand an overall platinum price deflation of more than 50 per cent.

TABLE 2.9. *Dufek Platinum Metals Mine relative profitability*

Factors flexed from +50 per cent to −50 per cent	Relative sensitivities		
	High		Low
Platinum price*	X		
Palladium price		X	
Pd/Pt ratio*		X	
Underground mine dilution*	X		
Head Grade (ROM GRADE)*	X		
Mill recovery	X		
Recoverable reserves			X
Underground mine rate	X		
Underground production	X		
Transportation, smelting and refining costs			X
Total operating costs*	X		
Annual personnel costs		X	
Annual supplies costs			X
Annual power costs			X
Annual transport costs*			X
Annual general and administration costs		X	
Total capital costs (CAPEX)	X		
Exploration capital costs			X
Shaft and equipment costs			X
Underground equipment costs			X
Concentrator plant costs		X	
General facilities costs			X
Project overhead costs			X
Working capital			X

*Plotted on sensitivity graphs, Models 1–16.

Perhaps the most interesting and valuable manner in which to evaluate the results is by restating one of the two questions posed earlier in this section; namely, what is the platinum concentration (grade) that one needs to find in the Dufek Massif in order to predict a viable or break-even venture (e.g. to produce a real IRR of 10 per cent)? The answers are listed in column 5, Table 2.10. This type of approach ricochets us back to the underlying assumption of platinum mineralization in the Dufek Massif, and whether or not such grades are realistic in geological terms. Let us therefore finish this chapter by addressing this aspect once again.

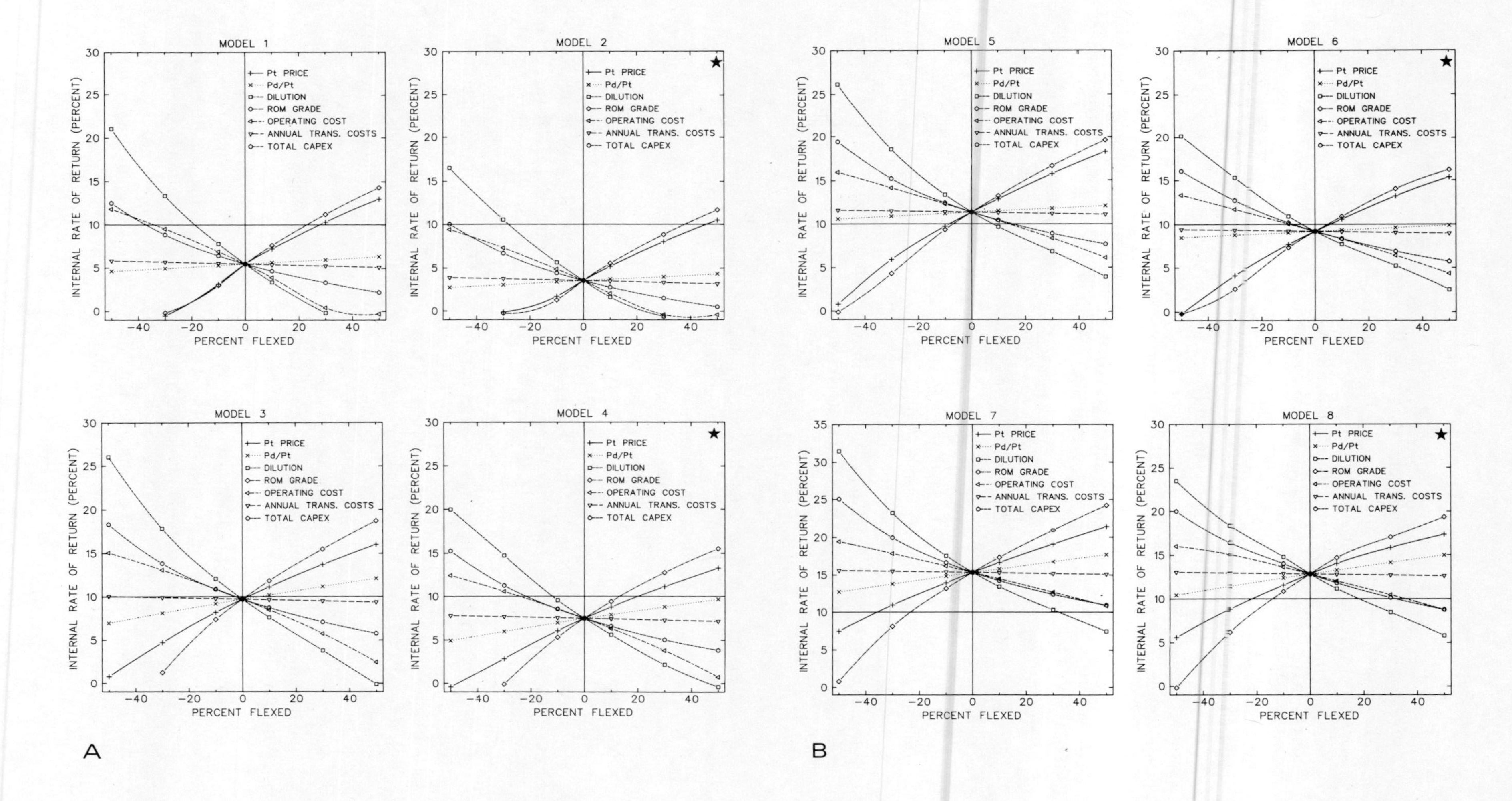
MODEL 1
MODEL 2
MODEL 3
MODEL 4
MODEL 5
MODEL 6
MODEL 7
MODEL 8
INTERNAL RATE OF RETURN (PERCENT)
PERCENT FLEXED
Pt PRICE
Pd/Pt
DILUTION
ROM GRADE
OPERATING COST
ANNUAL TRANS. COSTS
TOTAL CAPEX
A
B

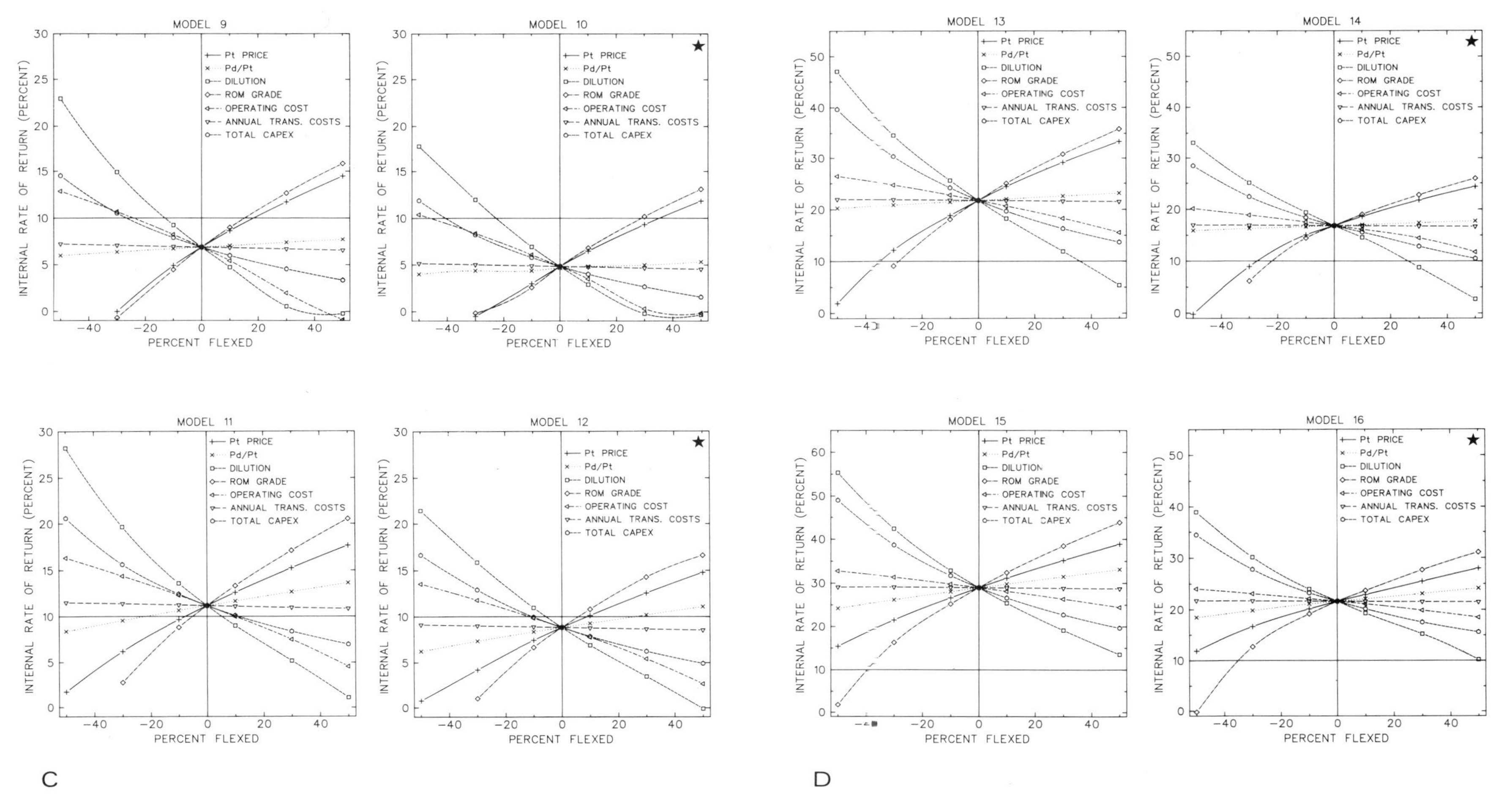

Fig. 2.11. Sixteen economic models with sensitivity analysis of the Dufek Platinum Metals Mine over the period 1986 to 2016. The details of the models are described in the text and summarized in Tables 2.8–2.10. Models have been divided into four groups (A–D) according to four specific platinum price paths. Within each group, the upper two models—e.g. Models 1 and 2 of Group A—have been calculated on the basis of a Pd/Pt of 0.42. The lower models have a Pd/Pt of 1.6. No taxation has been allowed for in the left hand models of each group—e.g. Models 1 and 3 of Group A. A World Bank taxation scheme has been incorporated in the right hand models of each group (indicated with a star). The horizontal reference line (=10 per cent real internal rate of return (IRR)) is the required discount rate set for the Dufek Platinum Metals Mine in this study. Below this rate, the project will not generate an acceptable economic return. The corresponding values of platinum grades (e.g. cut-off grades), mine dilution, changes in world platinum price, or changes the capital expenditure and operational costs of the Dufek Platinum Metals Mine that are needed to sustain this required real IRR value of 10 per cent in each of the models, are given in Table 2.10.

TABLE 2.10. *Dufek Platinum Metals Mine: break even values*

Economic model	Pd/Pt	Price Scenario	Change in Pt Price (%)	ROM grade (g/tonne)	Dilution (%)	Change in CAPEX (%)	Change in OPCOST (%)
Taxation–World Bank type							
2	0.42	A	+46	5.42	16	−50	<−50
6	0.42	B	+ 5	4.10	21	− 8	−9
10	0.42	C	+35	5.04	17	−40	−46
14	0.42	D	−27	3.07	28	>+50	>+50
4	1.6	A	+20	4.45	20	−22	−23
8	1.6	B	−23	3.39	26	+31	+33
12	1.6	C	+ 8	4.14	21	−10	−11
16	1.6	D	<−50	2.56	33	>+50	>+50
No taxation							
1	0.42	A	+28	4.85	18	−37	−34
5	0.42	B	− 9	3.57	24	+15	+14
9	0.42	C	+29	4.97	19	−28	−23
13	0.42	D	−35	2.84	30	+50	>+50
3	1.6	A	+ 2	3.98	22	− 3	− 3
7	1.6	B	−38	3.03	29	>+50	>+50
11	1.6	C	− 8	3.79	23	+12	+10
15	1.6	D	<−50	2.33	>34	>+50	>+50

Base case run of the mine (ROM) grade: 3.94 g/tonne platinum.
Base case mine dilution: 22.3 per cent.

Price scenarios for platinum: A. $475/ounce—constant
B. $475/ounce + 2 per cent p.a. real increase
C. $475/ounce, increased to $950 2 years pre-production followed by a 10 per cent real decrease for 6 years
D. $475/ounce, increased to $1520 2 years post-production followed by a 10 per cent p.a. real decrease for 10 years.

Direct comparison and contrasting of rock sequences has been one of the key building blocks of scientific advancement in geology. In theory, therefore, the first order similarities between the rocks of the Dufek and Bushveld complexes appear to carry significant weight in support of suggestions that the Dufek complex offers a unique target for platinum mineral exploration. On this basis, the ore grade used in the Dufek Platinum Metals mine models, or the projected ore values needed to mine profitably in the area, appear reasonable and are certainly not unduly speculative. Favourable platinum grades of much higher values have been found in layered igneous complexes of vastly different ages from various parts of the world, and indeed within the Bushveld complex itself, in layers other than the the Merensky Reef which was used in the Dufek models as a standard (tables 2.11 and 2.12). Nevertheless, the geological comparisons so far are by no means conclusive proof. The components for any potential mineral deposits make up only a tiny fraction of the total volume of liquid intruded to form a complex like the Dufek, and there are several reasons why they may or may not be concentrated in economically viable quantities. Four of these appear particularly pertinent.

1. The extent of homogeneity of the source region for layered intrusions, both in space and time. Thus, for example, the geological age of the intrusion has been suggested to be a determining factor as to whether or not economic mineral components will be present. Significantly, the Dufek intrusion is part of a much larger volcanic–plutonic igneous province, remnants of which occur on other continents such as South America and Africa (Fig. 2.12). These igneous rocks were all emplaced within a relatively short time span, during a period when these continents were still grouped together as one continent (Gondwanaland; see Chapter 5). It is in fact generally believed that these igneous rocks are related to the processes which broke-

TABLE 2.11. *Platinum group metals grade distribution in the Merensky Reef, Bushveld complex*

PGM	Proportion	Ore grade (g/tonne)	Head grade (g/tonne)	Plant recovery (78%)
Platinum	59.2	4.82	3.15	2.59
Palladium	25.2	2.04	1.34	1.10
Ruthenium	8.1	0.66	0.43	0.35
Rhodium	3.0	0.24	0.16	0.13
Iridium	1.0	0.08	0.05	0.04
Gold	3.2	0.26	0.17	0.14
Totals		8.10	5.30	4.35
Base metal		Weight (%)	Weight (%)	Weight (%)
Nickel		0.28	0.18	0.140
Copper		0.17	0.11	0.086

Source: Buchanan (1979*b*).

up this supercontinent. In southern Africa, a small (1200 km^2) intrusive complex, the Insizwa complex, the same age as and of similar composition to the Dufek complex, is known to contain platinum and copper enrichment in its basal sections (Tischler, Cawthorne, Kingston, and Maske 1981). At the time of the intrusion, the Dufek and Insizwa complexes were probably separated by less than 1500 km. (Fig. 2.12). Thus, the geological date does not appear to be in conflict with the prediction of potential ore-grade platinum mineralization in the Dufek complex. (See also Table 2.12.)

2. Variable crustal emplacement histories and related crystallization kinematics may be dominant factors which determine mineral concentration processes, especially as the parent magma of, for example, the Bushveld complex does not appear to constitute an exceptionally enriched PGE source (M. Tredoux, personal communication). Several eminent researchers of layered igneous complexes believe that multiple liquid injections and magma mixing during the formation and evolution of these large complexes may be the important factors controlling sulphide mineralization at well defined stratigraphic levels (e.g. Campbell, Naldrett, and Barnes 1983; Kruger and Marsh 1982; Davies and Tredoux 1985; Kruger, personal communications 1984). The relative effects of such

TABLE 2.12. *Relative proportions and reserves of platinum-group metals in selected ore-deposits from igneous complexes*

	Bushveld Complex South Africa						*Stillwater USA*		*Noril'sk USSR*		*Sudbury Canada*	
	Merensky reef		UG2		Plat. reef							
	P	R	P	R	P	R	P	R	P	R	P	R
Pt	59	333	42	437	42	160	19	7	25	50	38	3.4
Pd	25	141	35	365	46	175	66.5	23	71	142	40	3.6
Ru	8	45	12	125	4	15	4.0	1.4	1	2	2.9	<1
Rh	3	17	8	83	3	12	7.6	2.7	3	6	3.3	<1
Ir	1	6	2.3	245	0.8	3	2.4	<1			1.2	<1
Os	0.8	5			0.6	2					1.2	<1
Au	3.2	18	0.7	7	3.4	13	0.5	<1			13.5	<1.2
Total		565		1047		380		35		200		9
Grade	8.1		8.71		7–27		22.3		3.8		0.9	

P—grade (g/tonne).
R—reserves (millions of troy oz).
Source: Buchanan (1979*b*).

processes cannot be satisfactorily evaluated for the Dufek complex at present, but there is at least one observation that may be significant. It should be noted, for example, that the tie lines joining pyroxene pairs from the Dufek intrusion are somewhat shorter than pairs from the equivalent stratigraphic position in the Bushveld complex. This may reflect differences in cooling rates between the two intrusions, with the Dufek magma having cooled faster than the Bushveld magma. Much more field and laboratory observation is needed to substantiate the above, and furthermore, the possible effects of such cooling rate on mineral concentration processes are not known. For the present, the mineralogical and chemical similarlities between the Dufek and other complexes are so striking as to mitigate this point. Indeed, it now appears that the size of an intrusive complex is the overriding factor which determines whether or not platinum-rich sulphide mineralization will occur in a layered complex (Campbell *et al.,* 1983). The larger the complex, the more likely PGM ore-grade mineralization. The Dufek and the Bushveld complexes exceed the size of the Stillwater complex by more than an order of magnitude, which in turn is more than twice as large as the Insizwa complex. That augers well for potential platinum mineralization in the Dufek complex.

3. The chemistry of the host rock and the extent of contamination with the intruding liquid may vary and have an important influence on concentration processes of economic minerals.

4. Early influx of exogenous reducing fluids may be a prerequisite for economic grade sulphide-mineral precipitation during the early stages of intrusion of the complex (e.g. de Waal 1977). It has become clear for example that the presence or absence of ore-grade sulphides is closely related to early sulphur saturation of the melt (Skinner and Peck 1969; Buchanan 1978; Buchanan and Nolan 1979), but that nucleation of an immiscible sulphide phase in sulphur-saturated melts is inhibited at high values of oxygen fugacities (e.g. Haughton, Roeder, and Skinner 1974).

The last two points can be discussed together, since the origin of inferred reduced fluids associated with igneous intrusions is subject to speculation. One the one hand, they may be liberated from the envelope of sulphur- or carbon-bearing country rocks during contamination, inter-reaction and assimilation with the hot intruding mantle liquid (point 3 above). Such a mechanism appears to have played a significant role during the precipitation of nickel and platinum sulphide minerals in the Platreef zone of the Bushveld complex (D. L. Buchanan and Grunshaw, personal communications 1984; Table 2.12), as similarly speculated for the mineralization in the Norilsk complex in the USSR (Table 2.12) and the Insizwa complex. Significantly, the rocks into which the Dufek Massif was intruded offer great potential for similar conditions. In the south-east of the complex, for example. (Fig 2.4) these rocks are Paleozoic sediments which are known to contain limestones, coal deposits and graphite-bearing shales, xenoliths of which are believed to have contaminated the magma, at least locally (Himmelberg and Ford 1983). In fact, the high magnetic anomalies measured along the inferred boundaries of the Dufek complex may well coincide with near-surface mineralization of this type, as illustrated in Fig. 2.4. Alternatively, the reducing fluids may have emanated from the deep mantle during the intrusion of the complex, at which time they could have played a similar role of scavenging and precipitating metal components from the magma (point 4). Extensive flushing of large magma reservoirs by reduced volatiles is most favourable if the intrusion coincided with a period of prolonged normal magnetic polarity of the earth, since these periods coincide with times of increasingly reduced volative expulsion from the deep mantle to the earth's surface (Nicolaysen *et al.* 1982; Nicolaysen 1983). The age of the Dufek Massif indicates that it falls within such a phase of earth history. Indeed, thick sections of the exposed parts of the Dufek complex are normally magnetized, as are critical sections of the Bushveld complex, (Beck 1972; Hattingh 1983; Fig. 2.4), satisfying this requirement. It may, in fact, yet prove to be a truism that extended periods of reduced deep volatile expulsion and similar prolonged periods of a normal earth's magnetic field are closely related to volumetrically large-scale magma expulsion from the mantle at elevated rates. Both the Bushveld complex and the Dufek complex are part of much larger and extensive volcano-plutonic provinces and tectonic events at their respective times of intrusions, about 2000 and 180 million years ago (Fig. 2.12).

Manifestations of volatile fluid migration through igneous complexes are subtle, and the detection of their extent and effects are laborious to document. It requires, for example, stable

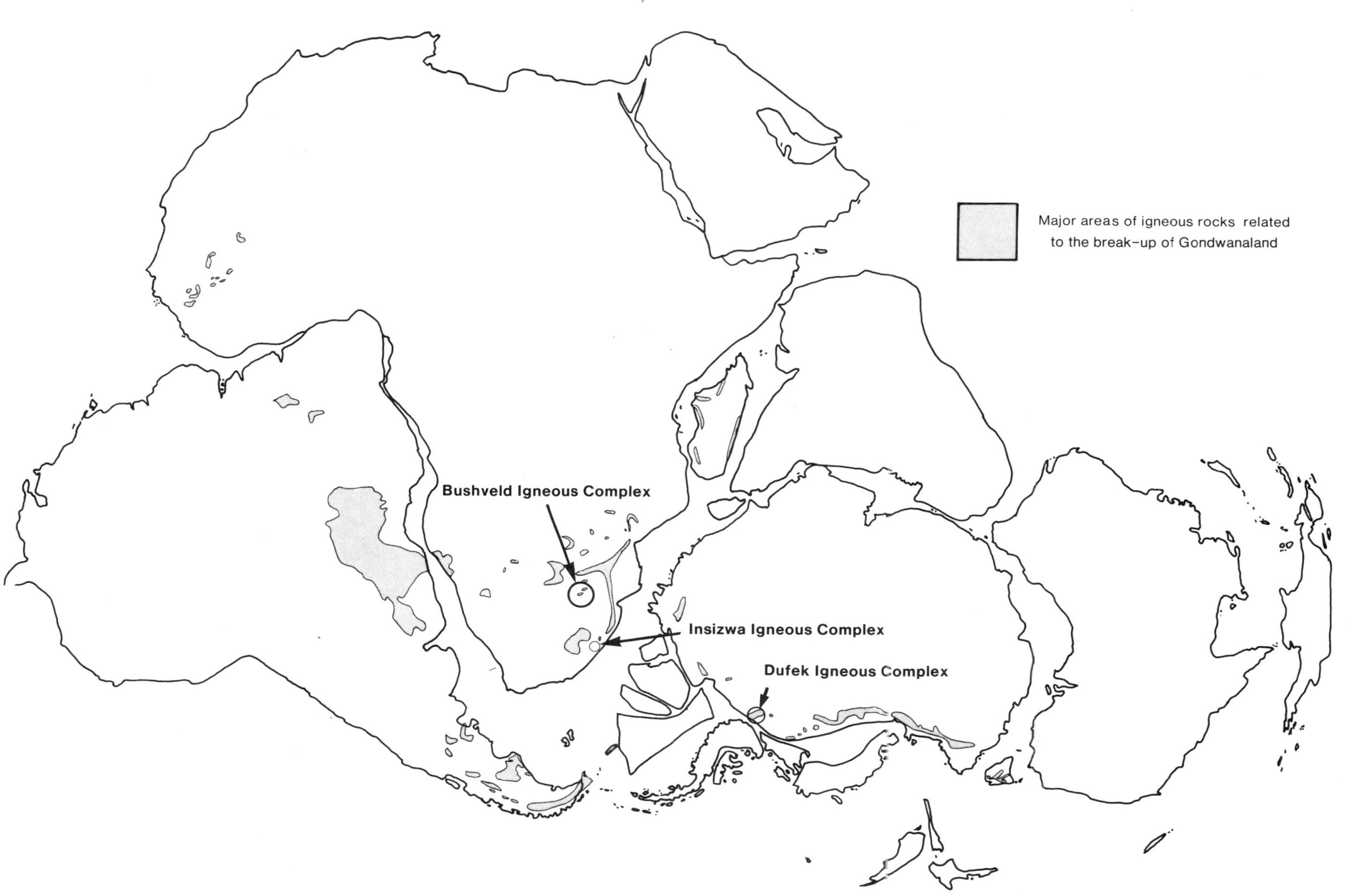

Fig. 2.12. Map showing the locations of three igneous complexes referred to in the text. Both the Dufek and Insizwa complexes were intruded into their respective continents when these continents were still part of a supercontinent Gondwanaland (Chapter 5). These two complexes were part of an extensive igneous provence, centred over a 'hot spot' region below the 'south-western' part of the Gondwanaland continent between about 200 and 120 million years ago. The Insizwa complex is known to have platinum mineralization. The Bushveld complex, which is also part of a much more extensive igneous province, invaded the crust of southern Africa 2000 million years ago. (After de Wit *et al.* 1985.)

isotope geochemistry. Such detailed chemical studies provide additional means by which the role of oxygen and sulphur fugacities in the production of an ore-rich sulphide phase in the Dufek-type magma can be predicted with greater certainty (Buchanan and Nolan 1979; Buchanan and Rowse 1984). Although there are difficulties in establishing reliable oxygen fugacity data, this has been successfully achieved by extrapolations for the Bushveld rocks, using magnetite–ilmenite pairs from the upper zone diorites of the Bushveld complex (Buchanan and Nolan 1979). Such tests have not been done for the Dufek complex, although preliminary investigations as well as other less elaborate geochemical tests that could help to predict platinum, nickel, and chromite mineralization more confidently (Cawthorn and McCarthy 1980) could easily be carried out using the systematically collected samples from the Dufek Massif, which are housed at the US Geological Survey in Menlo Park, California.

Summary and conclusions. The predicted mineralization of platinum group elements in the Dufek complex is based on valid geologic assumptions even though these should and can be more carefully constrained. The ore can be profitably extracted using 1983 polar technology and a reasonable set of economic parameters. In a broad sense, this is compatible with the predictions of Elliot (1976), who stated that mineral exploration of platinum from the Dufek Massif would not become profitable unless the market price of the metal increased by at least a factor of 2.9. Since his 1976 study, the platinum price has increased by a factor of 2.8. Additionally, the project as modelled in this study allows for significant economic optimization, using alternative, developing technology. Some of this is already being tested for polar utilization. For example, hovercraft technology may prove to be a much more cost-effective means of concentrate transportation. Ice transportation by hovercraft has been under test by the Japanese at their Syowa Antarctic base, using a Mitshui experimental hovercraft which was designed for ship-to-shore use and for inland transport over ice, with payloads of 600 kg over distances of 300 km (Antarctic 1981). Recent advances in hovercraft technology allow much greater payloads, of between 8 and 17 tonnes, speeds (80 km/h) and much improved efficiency (Hewish 1983).

There are many criticisms that can be levelled at specific aspects incorporated into the economic models of this mine. For example, the working capital used is probably too low, even though this is relatively insensitive to the overall profitability of the project. It is also unlikely that the mine and mill would be operated on the basis of one shift per day, especially given the mine's remote and inclement circumstances. Such shortcomings are further explored elsewhere (de Wit *et al.* 1984*b*). Additionally, as mentioned, the exploration period allowed for is short. Even within a relatively predictable layered environment, as in the Dufek complex, perhaps 5 years or more would be required to proceed rationally through the sequential information-gathering, exploration, delineation, test work, and evaluation stages.

On the other hand, we have incorporated the requirement for a 1000 m shaft. If the deposit is shallow, relatively flat lying with strike dimensions as the geology suggests, this capital expenditure would not be necessary. Moreover, the exercise has been one of evaluating the economics of developing a mine, rather than that of exploring for a potential opportunity. From this, it follows that mineral exploration for platinum ore-types in the Dufek Massif thus appears to be an attractive proposition today. *However, the bottom line of this exercise indicates that without exploratory drilling, speculations about future exploitation of these inferred Antarctic platinum riches will remain rife.* Financing the necessary drill holes and the comprehensive geochemical follow-up work to eliminate these uncertainties would cost in the region of US $5–10 million. Spread over a period of several years, and financed through international co-operation, this seems a small price to pay for such a potentially rich return from what, according to this study, could economically and environmentally be a relatively low-risk mining venture. It seems pertinent, therefore, to examine further now whether or not society at large might benefit from such an operation beyond the more short-term economic rewards. The next chapter will endeavour to consider this aspect.

3 THE STRATEGIC ROLE OF PLATINUM AS A RATIONALE FOR PLATINUM MINING IN ANTARCTICA

Apart from the technical advantages of mining a low volume high value element such as platinum in Antarctica, there are distinct economic, social, and political reasons for stimulating this experiment in the form of a platinum mine.

The six elements of the platinum group metals (PGM), namely platinum, palladium, iridium, osmium, rhodium, and ruthenium, are amongst the rarest and most expensive natural elements known. It is in part this rarity that has kept their economic behaviour relatively simple to follow and predict; the world market for PGM is small, well defined, steadily growing and extremely sensitive to supply and demand (Newman 1973, Buchanan 1979*a* and *b*, 1981, 1982). In fact, the scale of growth in Western world demand, despite a long-term upward price trend over the last 50 years, has been described as 'dramatic' (Johnson Matthey, in the House of Lords Report 1982; Table 3.1). In 1960, the total world production was estimated to be about 0.04 million kg. By 1977, this had increased to 0.2 million kg and in 1982, in spite of the general recession, total world production was approaching 0.25 million kg (Chaston 1982).

TABLE 3.1. *Western world platinum consumption (kg)*

1930	1940	1950	1960	1970	1980 (estimate)
5 000	18 000	11 500	21 000	42 000	90 000

Source: House of Lords (1982).

Total PGM demand growth rates are expected to range between 2 and 3.3 per cent per annum between 1985 and 2000 (Table 3.2). The USA and Japan together use about 80 per cent of the current world consumption; the EEC countries consume less than 10 per cent of the Western world's annual usage and possess within their borders no economic deposits of PGM. Nonetheless, the EEC, predominantly through the United Kingdom, controls a disproportionate position in the refining and marketing of platinum, in its recycling, and in the frontiers of research and development of its new uses in technology (Johnson Matthey 1982*a*).

The supply side of PGM could, in theory, easily be cartelized; almost 99 per cent of the world's PGM in 1980 was produced in three countries, namely the USSR (48.1 per cent), South Africa (45.2 per cent) and Canada (5.4 per cent; see also Table 3.3). Furthermore, only in the case of South Africa is mining conducted primarily for platinum; elsewhere it is produced as a by-product of base metal (notably nickel)

TABLE 3.2. *World primary demand growth rates (per cent increase p. a.)*

	1976–1985	1985–2000
Platinum	2.1–2.4	2.4–2.9
Palladium	2.0–2.2	3.1–3.9
Rhodium	5.5–6.9	1.8–2.0
Total PGM	2.0	3.3

Source: Glacken (1981).

TABLE 3.3. *Average new Platinum group metals supply to the non-communist world*

	Platinum (%)	Other platinum group metals (%)	Total platinum group metals (%)
Republic of South Africa	77	51	64
USSR	17	41	29
Canada	5	7	6
Others	1	1	1
	100	100	100

Source: House of Lords (1982).

exploitation which can severely constrain platinum output. Recently, for example, in the case of Canada, major nickel mines in the Sudbury area were forced into periodic closure to reduce output by almost 45 per cent in response to the depressed nickel market and increasingly stiffer competition from nickel production elsewhere. In consequence, PGM production fell by nearly 30 per cent in 1982 (*Mining Journal* 1983). In passing it is worth mentioning that China is self-sufficient in PGM production but that little is known about its further potential. Finally, in spite of strong opposition to mining the Stillwater complex because it occurs in the remote and environmentally protected area of Montana, USA, and because of technologically difficult mining conditions, it appears that a decision may soon be made to proceed with palladium-platinum exploitation from this complex (*Mining Engineering* 1984). If developed the deposit could supply most of the requirements of the United States for palladium and about a quarter of its platinum needs (Buchanan 1979 *a, b*).

The increasing demand for PGM is directly related to their inert chemistry. These noble metals have exceptional electrochemical characteristics and their catalytic properties are unique over a wide range of temperatures. The metals maintain their mechanical properties over long periods in oxidizing environments at high temperatures (Boswell 1982), and platinum-enriched alloys provide enhanced resistance to corrosion under agressive environmental conditions when compared to their conventional base-metal counterparts (Coupland, Hall, and McGill 1982). This ensures the platinum elements a crucial position in key industrial processes and, ironically, in the present urgent need for effective environmental protection from some of these same industrial processes.

In the last two decades, there has been encouraging progress in the utilization of platinum metals for cost-effective pollution control of harmful gaseous emissions, such as hydrocarbons (HC), carbon monoxide (CO), and oxides of nitrogen (NOx). For example, a world wide increase in the demand for nitrate fertilizers required a considerable increase in the use of rhodium–platinum catalyses for the oxidation of ammonia with air during the manufacture of nitric acid. More recently, these catalysts have been introduced for the control of nitrogen oxides formerly emitted from nitric acid plants as a characteristic brown plume. Above all, the greatest single use for platinum in the United States, amounting to over 57 per cent of the estimated annual consumption since the introduction of the US Clean Air Act Amendment bill of 1977, is for emission control catalysts for internal combustion engines. Similar increases for PGM demand can be expected in Europe and Australia, as legislation for motor-vehicle emission control and prohibitive lead contents in gasoline, similar to that in the USA and Japan, is implemented over the next 5–6 years (Herbert, Acres, and Hughes 1980; Diwell and Harrison 1981; ECE 1983). The most stringent emissions control regulations currently in force are those of the United States of America and of Japan. All cars marketed in those countries are fitted with noble metal catalysts; Australia will enforce similar regulations in 1985 (Johnson Matthey 1983*a*).

Significantly, PGM is also extensively used in the cracking and refining or petroleum to produce high quality (high octane), lead-free gasoline. Such lead-free petrol, without which conventional catalysts will not work because lead effectively 'poisons' these catalysts (and thus rapidly destroys their function), is not marketed in Europe (Johnson Matthey 1982*b*). Until it is, the ability of the motor industry in the EEC countries to meet more stringent emissions control standards will depend upon sophisticated mechanical engineering modifications and/or the development of catalysts whose effectiveness is not destroyed by lead. Lead-tolerant platinum-based catalysts are presently available, but their efficiency does not match those of the conventional catalysts, and they are ineffective in the control of NOx emission (Johnson Matthey 1982*b*). The UN Economic Commission on Europe (ECE) Regulation 15 prescribes the emissions standards enacted by most European countries. Amendments to improve these standards (e.g. ECE 15-04) are being introduced but any succeeding legislation has yet to be considered. This will be influenced by the deliberations of the EEC's *ad hoc* committee (ERGA; European Regulations Global Approach), but the decision-making process is slow.

Already the governments of Sweden and Switzerland have independently enacted emission standards representing the most stringent stage of non-catalytic control. Switzerland will enforce standards of still greater severity in

1986. These will be the equivalent of regulations established in the USA in 1977, for which the majority of cars were catalyst-equipped. However, Switzerland and most European countries have as yet expressed no intentions of introducing lead-free petrol, which would allow the use of conventional catalysts. The United Kingdom however, has declared its determination to remove lead from petrol by 1990.

Nevertheless, a more widespread European concensus will probably have to wait other influences. Mounting pressures arising from a more informed perception of environmental deterioration, increasingly recognized by most European nations, can be expected to dominate relevant decision-making processes. For example, introduction of legislation based upon lead-free fuel and so-called 'three-way' catalysts, which have proved in the USA and Japan to control simultaneously HC, CO, and NOx emissions, has been decided upon in the Federal Republic of Germany as from 1986. This is partly a result of scientific claims that NOx emissions from motor cars form a major contribution to 'acid-rain', that is causing damage to European forests, especially in Germany and Scandinavia. It is believed that Germany's decision will almost certainly be followed by many other European nations, and it is on this basis that recent estimates have predicted that Europe's annual platinum requirements could rise by 400 per cent by 1988 (Rand Daily Mail 1983), or by about 13–15 tonnes of platinum group metals annually (Emmel 1984). Such European consumption will not, however, be sustainable. Once catalyst-equipped cars reach the end of their useful lives, the indications are that it is possible to recycle cost-effectively the PGM, thereby eventually reducing the demand for new metal. This is the case today in the USA, where catalysts have been in use for 9 years.

In summary, future platinum demand curves related to the automobile industry can be predicted to follow paths of sharp increases followed by gentler exponential declines, as different industrial regions in the world implement rigorous emission control legislation.

Recent increases in fuel and feed stock prices, as well as pressures to conserve exhaustible natural resources, have also increased the use of platinum metals as catalysts in preference to base-metal catalysts, because of higher and cleaner yields. The fuel cell, using platinum during the catalystic oxidation of hydrogen, is now recognized as the largest potential user of platinum. Unlike a battery, the fuel cell is an engine which continues to generate electricity as long as fuel and oxidant are fed to it. Its application in the direct conversion of chemical energy into electric power was first advanced by the need for light-weight fuel cells in space projects, in which they continue to fulfill a role for space power generation and storage. Presently, it is predicted that a major requirement for platinum in the 1990s will develop as fuel cells become more widely adapted for small domestic industries and vehicle propulsions. Such applications are anticipated to repeat the success that platinum has had in the pollution control catalyst for the automobile industry (*Platinum Metals Review* 1983*a*).

Because fuel cells are currently some 30 per cent more efficient in energy consumption and because they are pollution-free reactors, they are expected to play a major part in the USA, perhaps as early as the mid-1980s. Fuel-cell power plants can be adapted to a wide range of operating loads, they respond quickly to load changes and have minimal requirements for water. These factors make installation of fuel cell power plants acceptable in environmentally sensitive urban areas close the point of energy requirement. Fuel-cell technology has already advanced to the point where 4.5 megawatt power stations are operating in New York and Tokyo (Cameron 1981; Philpott 1983; *Platinum Metals Review* 1984), and other multi-megawatt commercial plants are presently being planned (*Platinum Metals Review* 1983*b*). In the United States, a major programme of 'on site' testing of forty-nine 40 kW phosphoric acid fuel cells is underway. The first 40 kW fuel cell was started up in an Oregon laundry in 1982; the second fuel cell powers the facility of a telephone exchange for about 20 000 telephone customers in a small town in Connecticut. Both plants are connected to the local power grid to effect co-generation (Philpott 1983). Similar trials with 40 kW fuel cells commenced in Japan in 1982, and one plant has been operative since April 1983 (*Platinum Metals Review* 1983*b*). The Japanese Government is promoting a large research and development project on fuel cell power generation with the objective of reducing dependency on oil imports. Phosphoric acid and alkaline fuel cell activity has also been reported from Canada and several European countries (Philpott 1983; Van den Broeck and Cameron 1984).

If the initial trial-plants prove successful, the increased demand for PGM may be much as 10 tons a year (Glacken 1981). More accurately, at current electrode loadings (which are reported to be of the order of 10 kg of platinum per megawatt of installed capacity, with the possibility of fifty 27 MW fuel cell power plants being built in 1985 and up to 1460 plants in 1990) this represents an annual requirement of between 786 and 58 690 kg of platinum per year (Herbert *et al.* 1980; *Platinum Metals Review* 1983*b*).

Continued research in the United States, particularly at the NASA Len Research Centre, is focusing on new avenues for fuel cell application, mainly in gas and electricity supply utilities. Significantly for this study, major objectives include 500 MW annual capacity power plants for mining applications and power for remote settlements and installations, especially food processing plants in Third World countries (Philpott 1983). In Belgium, research into future electric vehicle developments has identified an opportunity to use city transportation as a first market for fuel cell traction power plants. Successful trials with mobile test beds, such as a Volkswagen van, are continuing and it is anticipated that rapid commercialization will follow once the system has been fully demonstrated (Van den Broeck and Cameron 1984).

Other more conventional uses of platinum alloys range from the fields of optical glass making and high potential performance glass fibre production, to their uses in the dental and medical profession (e.g. for cancer therapy), the textile industry, the jewellery industry (especially in Japan), the growing specialized electronics industries, and in electrochemical protection against corrosion. For example, cathodic protection of steel by coatings of platinum has led to the widespread development of systems for the protection of marine structures, rigs, vessels, oil-well casings, pipelines etc. (Hayfield 1983).

Thus it appears that the projected future for platinum looks bright. The 25th anniversary volume of *Platinum Metals Review* (1982) encapsulates it as follows:

> If any prophecy is safe, it is that catalyst for removing carbon monoxide, hydrocarbons and nitrogen oxides from exhaust gases of internal combustion engines will then have been overtaken by other major uses of platinum metals in the year 2007. In the motor industry, by then these catalysts may be incorporated in the combustion chamber to control pollution at source or new type of motive power may be in use, possibly electrically powered from fuel cells or external combustion engines such as the Stirling engine using a noble metal catalyst to burn the fuel cleanly and efficiently. Whether the largest use will then be for solar energy conversion or storage, in high strength super alloys, in advanced electronics systems or for one of the many potential applications at present being investigated in academic and industrial research establishments or even some yet unconsidered purpose, will become clear only with the passage of time.
>
> (Chaston 1982, p. 9.).

In summary, the expected increase in demand for platinum is closely tied to two very pertinent global socio-economic concerns, both presently under close scrutiny and being subjected to severe political pressures in the industrial world, namely energy conservation and, perhaps even more important, the curbing of pollution and improvement of environmental and health standards. On this basis, there appears to be sound justification for the optimism that 'we are at present (entering) in the age of the platinum metals catalyst' (Chaston 1982), and that the projected 2–3 per cent annual demand growth rates may well turn out to be conservative.

Given the reality of this future growth of the platinum market, are we now more clearly able to identify, and perhaps justify, the need to explore for these noble elements in such a remote, pristine, and fragile environment as Antarctica? Are there, in fact, specific socio-economic and socio-political arguments which could motivate exploitation in Antarctica for these metals and which might, in a cumulative manner, outweigh counter-arguments?

Whereas at present the PGM market appears economically at a general equilibrium, on the supply side there is a justifiable reason for major concern. South Africa contains the majority of the world's resources and reserves of PGM and dominates the **platinum** production (77 per cent) of which it has about 86 per cent of the world's reserves, whilst the USSR produces most of the world's **palladium**. Thus there is a distinct incentive for monopolistic pursuits. Indeed, the price paths of platinum since the late 1960s closely compare with those predicted from theoretical models of a monopoly within the framework of a small competitive fringe and market uncertainties (Dasgupta and Heal 1979; Fig. 3.1). Therefore, any decisions as to whether or not to develop additional capacity outside

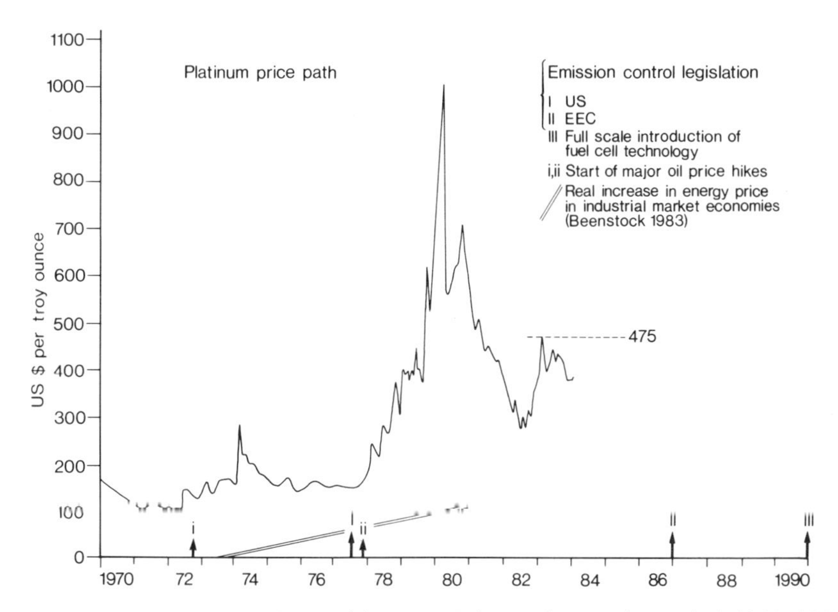

Fig. 3.1. Platinum price paths, expressed in US $/troy oz. Price path over the period 1969–1983. Also shown are the projected times, used in this study, for the start of emission control legislation in the EEC countries (II) and the full-scale introduction of fuel cell technology in the USA and Japan (III). Both of these events will require a substantial increase in the use of platinum group elements. The level of the platinum price used in this study (1983 US $475) is also shown. Sources: House of Lords (1982) and the *Weekly Mining Journal,* London.

South Africa to meet a future demand for platinum must take into careful consideration South Africa's position as a key producer of this metal. Failure to do so will ultimately have detrimental consequences both in the form of creating unnecessary welfare penalties and in foregoing potential social welfare rewards.

Firstly, it is well known from theory and observations that monopolies do not set mineral price paths to optimize long term welfare (Hotelling 1931; Dasgupta and Heal 1979). A new large supply of platinum on the world market could disintegrate this monopoly framework, thus drastically reducing world prices (*cf.* Robson 1979; Dasgupta and Heal 1979), in a way comparable to that proposed in a model of monopoly disintegration within OPEC (Beenstock 1983).

More intuitively, the mere realization of a larger world stock of platinum, which in the case of the Dufek Massif might conceivably increase world stocks by as much as 100 per cent, will simply decrease the spot prices and reset long term price paths in a predictable manner (Nordhaus 1973; Arrow and Chang 1978; Robson 1979). Although the fine details of such a price path will vary due to inherent uncertainties associated with the new deposits (Hartwick 1983), there is nevertheless clear evidence from the carefully balanced platinum producer price and profit margins, production rates, and market price and requirements (Buchanan 1979*a*, *b*; Johnson Matthey 1982*a*; Waddell 1982; Christian 1983) that such a major hypothetical shock adjustment would be absorbed in South Africa by a fall in production and massive redundancies. Such a scenario could create immense social disorder; the mining industry is South Africa's single largest employer of labour, the overwhelming majority of which are unskilled workers (Neethling 1983). On the other hand, the vast and easily accessible South African platinum (*cf.* Newman 1973) and cheap labour resources could counteract any attempts to exploit a remote Antarctic deposit. A relative small decrease in the price of the platinum group elements would assure the failure of such a venture. The profit margins of the three South African platinum producers—two of which control an internal duopoly—are large (Newman 1973; Fig. 2.5). Thus, on economic grounds, the predicted price kinematics would most certainly force any mining venture in Antarctica to be shelved and would reaffirm the South African monopoly, albeit at a lower profit margin. This would have the added negative consequence of knowing that global platinum resources are not being optimally exploited, because theoretically such optimization can only be approached with increasing knowledge of the global finite resources (Robson 1979; Dasgupta and Heal 1979, Hartwick 1983).

In the light of the economic and technical considerations *vis à vis* the intrinsic properties of platinum, and given the geological constraints of the world's non-uniform endowment of PGM, it is not surprising that politically these noble metals are increasingly being classified as strategic minerals in the Western industrialized nations of the OECD (van Rensburg and Pretorius 1977; Pretorius 1979; Chamber of Mines of South Africa 1980; US Senate Committee on Foreign Relations 1980; Vale 1980; Buchanan 1982; Waddell 1982; House of Lords 1982; Hansard 1983; Neethling 1983; see Andor (1985) for overview). South Africa, as the monopoly producer, continues to exploit this knowledge and 'advertises' so openly, arguably using it politically to put leverage on industrial platinum buyers towards mutually beneficial trade agreements (Vale 1980; Geldenhuys 1981).

Thus, apart from clear economic reasons for OECD countries to search for new platinum resources, there is ample political incentive to increase such prospecting momentum. In addition, there are social and moral obligations to look elsewhere. Foremost, the OECD countries need to free themselves from their Catch-22 trap, to enable them to separate their socio-economic demand for platinum from an apparent stake in the South African political status quo with its unacceptable structure of institutionalized racial domination in the form of apartheid policies. Conversely, but more bluntly, they need to assure for themselves an uninterrupted supply of this strategic mineral in case of sudden eruption of internal political upheaval in South Africa. The latter is projected by many as a realistic future scenario (McNamara 1982). Clearly, Antarctic platinum would not only offer a unique solution, but also, as a corollary, a new large external platinum supply could, if skilfully exploited, be used as an economic lever on South Africa in exchange for socio-political action to correct its human rights violations. In theory, this could be achieved whilst maintaining reasonable socio-economic stability.

In conclusion, it appears that aside from increasing economic welfare, the mining of Antarctic platinum could in theory be used more directly in helping to solve a number of global needs. Examples are:

more efficient energy production and conservation.
greater optimization of socially acceptable living standards in today's ever increasingly polluted environments.
greater and more efficient use of the world's exhaustible platinum resources through enforcing negotiated national and international exploitation.
amelioration of human rights violations.

In a world of potential social and economic disorder, such a challenge will require courageous political and managerial skills of weighing up the risks associated with mining Antarctic platinum, against the long term benefits. How can this be best approached? Can we afford to avoid confronting the issue, facing the risk of further social upheavals? If not, how can the risks be better quantified and the benefits be best defined in order to optimize decision-making? Can an equitable framework be established for a stable management of Antarctica's resources? Can the principals as outlined for platinum be extended to Antarctica's other exhaustible resources? In the concluding chapters of this study it is intended to focus on, and penetrate more deeply into, these questions.

4 A FUTURE FOR ANTARCTIC MINERAL RESOURCES

Antarctica is unique, an entire continent of disputed territory.

(Auburn 1982)

There can be no justification for the exploitation of Antarctica, except in terms of human greed. For we do not need Antarctica's supposed resources—we merely desire them to prolong a way of life which must, ultimately, come to terms with its own bankruptcy.

(Friends of the Earth 1982)

Productive employment opportunities must be found not only for the approximately 300 million people at present unemployed or inadequately employed, but for a total of 1000 million if those who will be entering the employment markets of the Third World over next 25 years are included. The pressure on natural resources, including land, and on the environment will further intensify and may constitute additional obstacles to economic growth.

(Blanchard, Director General ILO 1976).

But as I have pointed out, given today's level of complacency in some quarters, and discouragement in others, the more likely scenario is a world (population) stabilized at about 11 billion. We call it stabilized, but what kind of stability would be possible? Can we assume that the levels of poverty, hunger, stress, crowding, and frustration that such a situation could cause in the developing nations—which by then would contain 9 out of every 10 human beings on earth—would be likely to assure social stability? Or political stability? Or, for that matter, military stability? It is not a world that anyone wants.

(McNamara 1977).

4.1. AN EVOLVING ANTARCTIC TREATY AND THE EMERGING CONCEPTS OF COMMON HERITAGE

4.1.1. The Antarctic Treaty and progress towards establishing a minerals management regime

The termination, in 1991, of the first phase of the Antarctic Treaty will be a milestone of a very successful 30 years of Antarctic science and politics. Numerous statements by representatives of world communities have by and large expressed satisfaction, and indeed praise, for an almost perfect record of guardianship of Antarctica by the 12 founder-members and the four subsequently adopted members of the consultative circle of the Antarctic Treaty (Table 4.1 and Appendix A1). This success is almost entirely due to the emergence of a cooperative bond between an international group of Antarctic scientists during and preceding the International Geophysical Year (IGY) in 1957–1958, and which led directly to the birth of the Antarctic Treaty in 1961. This long period of international scientific co-operation and development in Antarctica, together with the peaceful management of Antarctic affairs, is all the more remarkable when projected against the background of the world's ongoing cold-war. The farsightedness of the scientists responsible for the formulation of IGY has thus made it possible to assess the forces and strengths of social and political stability in a framework of international scientific collaboration.

With hindsight, the conception, birth, and subsequent development of this Treaty has proved to be a major achievement of negotiation amongst a diverse group of nations with highly variable initial stakes and trumps (Appendices A1, A2; Auburn 1982; Bush 1982; Quigg 1983). For example, of the initial 12 founder members, seven claim sovereignty rights over various parts of Antarctica, three of which overlap considerably (Fig. 4.1; Table 4.1). Sovereignty claims, all established before IGY, are upheld on grounds which range from contiguity, discovery or recognition to purported effective occupation by explorers, scientists and military personnel (Auburn 1982, Appendix A2). With the exception of the area claimed by Norway, the claims—in accordance with the 'sector' principle—are all pie-shaped, their apices meeting at the South Pole. The undefined latitudinal boundaries of Norway's claim (Fig. 4.1) are assumed to be intentionally ambiguous because the 'sector' principle is rejected by Norway in Arctic polar territory. The 'left-over' area to the south of Norway's claim, together with a large unclaimed chunk of Mary-Bird Land, are today effectively *res nullius* (i.e. the property of no-one and, consequently potentially subject to the claims or exercises of ownership or sovereignty

TABLE 4.1. *Overview of Antarctic Treaty membership and Antarctic claims*

Antarctic Treaty membership—31 total (July 1984)

Consultative members—inner circle, with voting rights (16)				Observational members (15) (Status since 1983)
Founder members (12)				
Claimants (7) (*)	Sectors claimed (% of 360°)	Non-claimants (5)	Adopted members (4) (with date of adoption)	Acceded members (14) (with date of accession) Succeeded members (1)§
Argentina (1943)	25°W–74°W (14%)	Belgium	Brazil (1983)	← Brazil (1975)
Australia (1933)	45°E–135°E and 142°E–160°E (30%)	Japan	F.R.G. (1981)	← Federal Rep. Germany (1979)
Chile (1940)	90°W–53°W (10%)	South Africa	India (1983)	← India (1983)
France (1924)	136°E–142°E (2%)	U.S.S.R.	Poland (1977)	← Poland (1961)
New Zealand (1923)	160°E–150°W (14%)	U.S.A.		Bulgaria (1978)
Norway (1939)	20°W–45°E (18%)			Czechoslovakia (1962)
United Kingdom (1908)	80°W–20°W (17%)			Denmark (1965)
Total area claimed	(83%)			Finland (1984)
Overlapping claims				German Democratic Rep. (1974)
Chile–UK	80°W–53°W (8%)			Hungary (1984)
Argentina–UK	74°W–25°W (14%)			Italy (1981)
Chile–Arg.–UK	74°W–53°W (6%)			Netherlands (1967)
Total area disputed	80°W–25°W (15%)			Papua New Guinea (1981)
Unclaimed territory	150°W–90°W (17%)			Peru (1981)
				Rumania (1971)
				Peoples Rep. China (1983)
				Spain (1982)
				Sweden (1984)
				Uruguay (1980)

International composition of Antarctic Treaty membership†
(No. of countries; % of total)

		Voting members	Total members
United Nations		16; 100%	31; 100%
OECD		9; 56%	13; 48%
COMECON (CMEA)		2; 13%	7; 23%
Third world			
\$230	Average GNP per capita (US\$ 1979)‡	1; 6%	2; 7%
\$477	Average GNP per capita (US\$ 1979)‡	1; 6%	3; 10%
\$1590	Average GNP per capita (US\$ 1979)‡	5; 31%	10; 32%
Group 77		4; 25%	8; 26%
Non-aligned		2; 13%	3; 10%
EEC		4; 25%	7; 23%
OAS		4; 25%	6; 19%
OPEC and OAU		0	0

Antarctic treaty: Signed 1 December 1959; entry into force 23 June 1961. Ratified by all 12 original signatories.
Area of application: South of 60°S. Latitude, including ice shelves, without prejudice to international high seas rights.
Acronyms: COMECON—Council for Mutual Economic Assistance–planned economic development; EEC—European Economic Community; OAS—Organization for American States; OPEC—Organization of Petroleum Exporting Countries; OAU—Organization of African Unity; OECD—Organization for Economic Co-Operation and Development.
*Date of claims consolidated.
†Sources: *Europa Year Book* (1983); Union of International Associations (1984).
‡Source: World Bank (1982*b*).
§Papua New Guinea (1981).

by states, based on occupation, though many international lawyers now believe that no further claim to exclusive sovereignty over any portion of Antarctica would be recognized today).

It is open to speculation why other nations with confirmed long-term Antarctic interests, such as the USA or the USSR, have neither claimed this substantial remaining slice, nor on arguably legitimate grounds challenged other claims. There is a consensus of belief that this is related to an admixture of an acknowledgement of having missed the boat, and a recognition that the ambiguity of claims could be to their advantage. On the other hand, their refusal to recognize sovereignty rights over any parts of Antarctica benefits them by enabling them to retain a legitimate interest in the entire area.

Antarctic Treaty rules are careful to protect pre-existing claims, whilst simultaneously catering for those Treaty members that do not recognize these claims, through effectively freezing discussions on this topic for the period during which the Treaty is effective. Thus, until 1991, when the Treaty will be subject to potential renegotiation, the signatories will have been able to avoid sovereignty issues. A spin-off of this arrangement has been the effective keeping of Antarctica almost free from military confrontations, nuclear experimentation, and environmental pollution. Antarctica is a continent where peaceful scientific endeavours are protecting it from significant environmental damage and successfully buffering it from major political confrontations.

For example, whilst Chile and Argentina continue to argue vigorously over border disputes in southern South America, and whilst the United Kingdom and Argentina have been at war over the sovereignty rights of the Falkland Islands/Islas Malvinas and South Georgia Island, scientific negotiations at Antarctica meetings have always continued on a friendly basis, even **during** the recent war (J. Heap, personal communication 1983; Tucker 1983*a*). This amiability persists despite the fact that it is these three countries that are at loggerheads over sovereignty rights to the same territory on the Antarctic Peninsula (Fig. 4.1; Table 4.1), and that the abovementioned disputed areas and islands in the southern oceans serve as strategic stepping stones for supply and access to these overlapping Antarctic claims.

What is the explanation for this paradox? The strength of the Antarctic Treaty? Co-operation and loyalty amongst the Treaty nations is unquestionable, as reflected in their strict adherence to the Treaty's regulations (Auburn 1982; Quigg 1983); although Auburn (1984) and ASOC (1984) quote serious breaches of the Treaty, including assertion of sovereignty while the Treaty is in force; failure to communicate required information, and failure to adhere to the Agreed Measures for the Protection of Antarctic Fauna and Flora. An example of close adherence to the Treaty, is their firm stand on the prohibition of nuclear waste disposal. The only known Antarctic nuclear power station was built under supervision of the United States military wings at the scientific colony McMurdo in 1961. Following reports of potentially dangerous damage to the unit, it was (under US Naval supervision) dismantled in 1973–1974 and, together with any contaminated indigenous soil and foundations, shipped back to the US by 1979, (Naval Nuclear Power Unit, Fort Belvoir unpublished report 1973; *Antarctic Journal* 1980). It is therefore undeniable that forces mobilized through the Treaty obligations have given the Antarctic Treaty 'club' a unique flavour of successful international co-operation.

The ultimate strength and flexibility of the Treaty, however, is yet to be tested. Minerals activities in Antarctica are at a minimum, arguably because the Treaty nations agreed, in 1977, to an informal moratorium, or policy of voluntary restraint on such activities. This is conditional on adequate progress being made towards an acceptable minerals regime. The need, therefore, to formulate a pragmatic minerals regime before the potential renegotiation of the Antarctic Treaty in 1991 has become the most urgent task, but it is one which cannot be negotiated in the same secrecy as some of its predecessors, such as the CCAMLR or the conservation of Antarctic seals, have been. The Antarctic minerals debate has become a public issue with emotional overtones because of the enormous real social and moral consequences looming behind the potential costs, risks, and returns. The debate has, moreover, recently become open, following the discussion of Antarctica—for the first time ever—by the United Nations' General Assembly at its 1983 session (November 1983; Beck 1984). The result is that the Treaty 'club' must now face the issues squarely. To ensure either a worthy successor to the present Antarctic Treaty or a revitalized version thereof, it is vital that newly appreciated socio-economic ramifications should also be taken into consideration. Unfortunately by all accounts this is being diplomatically resisted or avoided (e.g. Tonge 1983; *Nature* 1983*c*). Why is this so?

At the time of IGY, very little was known about any mineral resource potential in Antarctica. As a consequence of the subsequent concerted scientific undertaking, (see for example Wilson 1961), by the time the Treaty

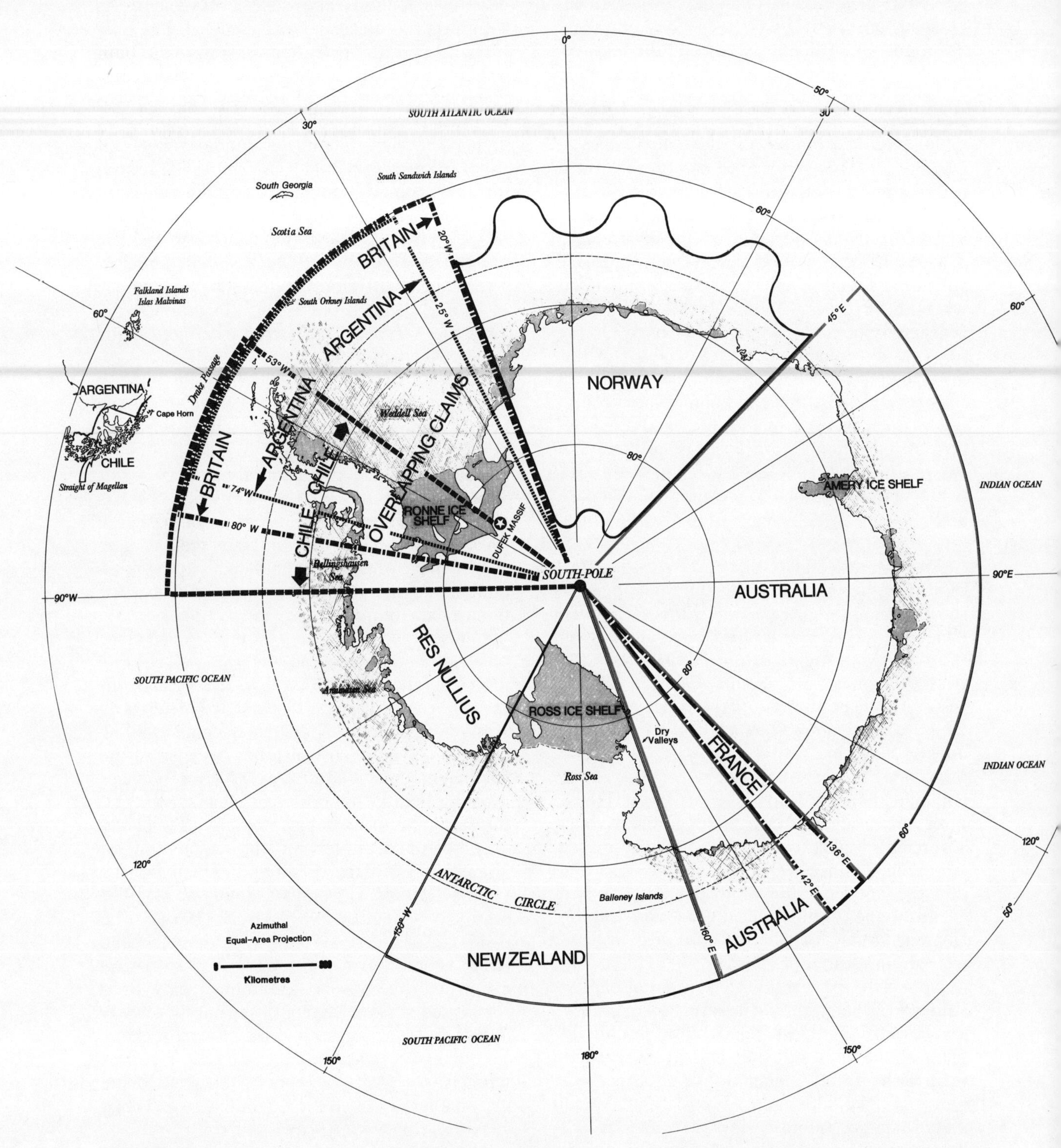

Fig. 4.1. Antarctica, showing the seven territorial claims, three of which overlap considerably. Note that both the inferred 'mineral storehouse' of the Dufek Massif and parts of the potential oil-rich Weddell Sea lie in the claimed sectors of Britain, Argentina and Chile.

was being negotiated the Dufek Massif was known to exist (Neuberg, Thiel, Walker, Behrendt, and Aughenbough 1959). Assuming its mineral potential was appreciated, the topic must have been deliberately circumvented, for it is the general opinion of Antarctic political scholars that the sovereignty issues could not have been avoided in formulating the Treaty if the convention had also attempted to govern mineral resources and provide rules for exploration–exploitation leases, royalties and so on. Thus, to date, there is no provision in the Treaty expressly regulating mineral policies. However, during the lifespan of the Treaty, its guardians have spelled out a number of fundamental principles, designed to promote the prudent management of Antarctica's non-renewable resources.

A start to this was made at the eighth consultative meeting in Oslo (1975) when SCAR was asked to assess the possible consequences of exploration and exploitation in Antarctica. This initiative eventually led to the establishment in 1976 of a SCAR group of specialists on the Environmental Impact Assessment of Mineral Resources Exploration–Exploitation in Antarctica (EAMREA, changed in 1981 to AEIMEE —Antarctic Environmental Implications of possible Mineral Exploration–Exploitation), which submitted its report to the ninth consultative 'club' meeting in London in 1977 (Zumberge 1979*a*). In addition to numerous official 'club' and SCAR reports, there have been private statements, specifically concerning mineral resources and the environmental consequences of their exploitation (Elliot 1976; Holdgate and Tinker 1979; Auburn 1982; Bush 1982; US Department of State 1982; Mitchell 1983; AEIMEE 1981, 1983; Holdgate 1983*b*; *ECO* 1983, 1984*a, b*), but serious negotiations amongst the Treaty nations towards establishing a framework for an acceptable minerals regime did not get under way until their eleventh consultative meeting in Buenos Aires in 1981. On this occasion, however, the signatories officially drew up guidelines recommending, *inter alia*, to their governments that 'a regime on Antarctic mineral resources should be conducted as a matter of urgency'. The most important reason for the urgency is that the issue of minerals in Antarctica, if left unresolved, may present a threat to the Antarctic Treaty and its regulatory system. This is because the issue is likely to bring back to centre stage the disputes as to sovereignty which the Antarctic Treaty so successfully avoids (Beeby 1983; Auburn 1984).

A scheme ('the Beeby draft') for the establishment of a minerals regime was subsequently drawn up and approved at the first session of a special consultative meeting on mineral resources in Wellington, New Zealand (June 1982; Bush 1982; Beeby 1983; Auburn 1984). Since then an informal seminar on this subject, held in Antarctica at the Chilean Air Force Base 'Teniente R. Marsh', (King George Island, South Shetlands, October 1982; Holdgate 1983*a*; Pinochet de la Barra 1982; Infante Caffi 1983; Vicuña 1983) has been followed by a more formal meeting in New Zealand, (Wellington, January 1983; SCAR 1983) and a second session of a special consultative meeting on mineral resources in West Germany (Bonn, July 1983; *ECO* 1983). Subsequent consultative meetings to further consolidate a minerals regime have been held in Washington (January 1984; *ECO* 1984*a*), and Tokyo, May 1984 (*ECO* 1984*b*). A further negotiating session is set for Brazil in February 1985.

It is believed that at the second session held in Bonn in 1983, the Treaty members proposed the establishment of a regime to govern exploration, in the form of a Convention which, whilst continuing to circumvent sovereignty issues, will permit prospecting and extraction on applications by the Treaty nations or by countries which the Treaty 'club' agrees to sponsor (*ECO* 1983; Beeby 1983; Auburn 1984). Such a regime—premised on side-stepping the sovereignty question—now appears to have been accepted in principle as a working model (*ECO* 1984). The proposed convention is likely to specify environmental protection standards, lay down conditions of access by operators, and provide for a licensing authority and approporiate revenue collecting and enforcement agencies. There will also be some kind of commission to monitor the success of the machinery thus established, with the power to review and amend as required (Brennan 1983*b*; *ECO* 1984*a, b*).

There is no evidence to suggest that the increasing concern of the Treaty members to develop a minerals regime can be attributed to long-term planning on their part. More probable is that various external pressures provided the principal impetus. The first of these is the fact that, by the early seventies, the world was seriously confronted with the importance of natural resources, their possible future shortages and their strategic values. The Club of Rome report, 'The Limits of Growth', was published in 1972 (Meadows, Randers, and Behrens 1972). This was rapidly followed by the sharp rise in oil prices in the wake of the first oil crisis in 1973. International awareness of the need for secure sources of strategic minerals intensified from this time on, as evidenced by the OECD and Brandt reports (OECD 1979; Brandt 1980, 1983). Secondly, by the early eighties, independent institutions and private individuals had publicly stressed the need for a minerals regime and published specific models to, *inter alia,* ensure environmental protection in the face of potential Antarctic exploitation (e.g. Elliot 1976; Alexander 1978; Pinto 1978; Rich 1982; Mitchell 1982; International Institute for Environment and Development (IIED) and International Union for Conservation of Nature and Natural Resources (IUCN)—See Mitchell 1983 and Kimball 1983*a, b* for chronological details). Thirdly, and even more importantly, the lengthy United Nations discussions on the 'common heritage of mankind', especially in the context of the management of the world's oceans, has had a profound influence on decision-making at Treaty 'club' meetings.

From its initial limited ambit (focused especially on the Deep Sea Bed), the concept of the 'common heritage of mankind' has won increasing recognition amongst the international community, and is gradually being invoked in an increasing range of circumstances. It has also now expressly been applied to Antarctica, as evidenced by the call of the representatives of Sri Lanka, at a meeting of the United Nations Conference on the Law of the Sea, for the 'common heritage' principle to be applied to Antarctica as well. This suggestion, echoed by the president of the Law of the Sea conference, H. S. Amerasinghe, was made in October 1975; and was reiterated by C. Pinto in July 1977 (Mitchell 1983; Quigg 1983). The statement of principles (in Articles 136 to 149)—introducing Part XI of the Convention of the Law of the Sea 1982—underlines the importance of the 'common heritage' approach and evidences the international recognition the concept has now been accorded. This lends immeasurable force to the call by non-Antarctic Treaty nations for the same principle to be applied to Antarctica. By 1980, the impact of these developments on the Antarctic Treaty 'club' was clearly evident, for the 'club' then pronounced itself "mindful of the negotiations that are taking place in the Third United Nations Conference on the Law of the Sea" (eleventh consultative meeting, recommendations XI-1, Appendix A3). Moreover, other evidence of assimilation of 'Law of the Sea' concepts in the present negotiations for an Antarctic minerals regime may readily be perceived (Vicuña 1983). The reason for this—by all accounts—is to counter criticism levelled at the Antarctic Treaty System as an exclusive 'club'; and to help preclude Antarctic Treaty Consultative Parties from losing control of the Antarctic and its mineral riches to wider public decision-making and management (Beeby 1983; Auburn 1984; Beck 1984). Recently, at least one consultative party (the United Kingdom) has publicly expressed its opposition to such a development (Heap 1983, at the United Nations General Assembly).

Bold initiatives to include Antarctica in 'common heritage' discussions within the United Nations General Assembly were made by India as far back as 1956 (Bush 1982). The negative response to such requests by Treaty-bound members at the UN (as documented by Bush 1982) was initially successful in diffusing attempts to draw Antarctica directly into this arena: but the situation seems now to have changed, for two main reasons. Firstly, the nature of the United Nations has altered since 1956 and the time (1958—1961) when the Antarctic Treaty was being formulated. At that period, the United Nations was still dominated by the Western industrialized nations, especially the United States. Today, it is the domain of the more numerous Third World nations. Secondly, during this transition period of power redistribution, the major occupation of the United Nations as regards the common heritage concept became the cumbersome and protracted Law of the Sea negotiations (UNCLOS). As a result, the principal interests and energies of the least developed countries over the last decade have been directed towards guiding the UNCLOS III deliberations to a satisfactory conclusion. This has now been largely achieved (Goldstein 1982, 1983; *Mining Journal* 1982; Brown 1983, Archer 1983); it is therefore not surprising that United Nations' attention is once again being directed towards Antarctica as part of the 'common heritage of mankind'. This time, however, the 'common heritage' view of Antarctica has been widely accepted, as reflected in the following address to the 37th Session of the United Nations General Assembly:

> ... there are still land areas which have neither natives nor settlers. There is, therefore, no one to inherit the land and to set up viable governments, should the claims of the metropolitan powers be given up. Because of this, little attention has been paid to these areas.
>
> It is now time that the United Nations focus its attention on these areas, the largest of which is the continent of Antarctica. A number of countries have in the past sent expeditions which have not limited themselves to mere scientific exploration, but have gone on to claim huge wedges of Antarctica for their countries. These countries are not depriving any natives of their lands. They are therefore, not required to decolonize. But the fact still remains that these uninhabited lands do not legally belong to the discoverers as much as the colonial territories do not belong to the colonial powers.
>
> Like the seas and the sea beds, these uninhabited lands belong to the international community. The countries presently claiming them must give them up so that either the United Nations administer these lands or the present occupants act as trustees for the nations of the world.
>
> Presently, exploitation of the resources in Antarctica is too costly and the technology is not yet available. But no doubt, the day will come when Antarctica can provide the world with food and other resources for its development. It is only right that such exploitation should benefit the poor nations as much as the rich.
>
> Now that we have reached agreement on the Law of the Sea, the United Nations must convene a meeting in order to define the problem of uninhabited lands, whether claimed or unclaimed, and to determine the rights of all nations to these lands. We are aware of the Treaty of Antarctica concluded by a few nations which provides for their cooperation for scientific research and prohibits non-peaceful activities. While there is some merit in this Treaty, it is nevertheless an agreement between a select group of countries and does not reflect the true feelings of members of the United Nations or their just claims. A new international agreement is required so that historical episodes are not made into facts to substantiate claims.

(Part of an address by Malayasian Prime Minister at the 37th Session of the United Nations General Assembly, September 1982).

Assuming, as seems to be the case, that this address reflects the consensus of opinion amongst the majority of member states of the United Nations, it may be legitimately queried whether the present Antarctic Treaty has become anachronistic.

The majority of nations—which are not bound by the Treaty—clearly regard the United Nations (with its emphasis on the 'common heritage' concept) as the appropriate negotiating forum for formulating a new regime to govern Antarctic affairs. This was reaffirmed at the Non-Aligned Movement[1] meeting in New Delhi (March 1983), through its communique which (*inter alia*) states:

> The heads of State or Government, while noting that relevant provisions of the Antarctic Treaty of 1959 related to international cooperation in Antarctica, in view of the increasing international interest in Antarctica, considered that the United Nations at the 38th Session of the General Assembly, should undertake a comprehensive study on Antarctica, taking into account all relevant factors, including the Antarctic Treaty, with a view to widening international cooperation in the area.
>
> (Antarctica: Memorandum 1983)

The fact that Antarctica was, for the first time ever, included on the United Nations General Assembly agenda for discussion in November 1983, underlines the significance of this Third World impetus and the strength of feeling upon which it is predicated. The General Assembly debate which followed in due course was noteworthy for the number of countries (40) which participated in the discussion. The resolution finally adopted by the Assembly called for the preparation by the UN Secretary-General of a comprehensive, factual, and objective study on all aspects of Antarctica, to be further discussed at the 39th session of the General Assembly, in November 1984. Almost all the consultative parties to the Antarctic Treaty supported such a study; and an item entitled 'The Question of Antarctica' has been included in the provisional agenda for this 39th session (UN General Assembly Agenda item 140, 1983).

These developments herald the possible beginning of a major reassessment of the Antarctic Treaty and its ability to meet the challenges that lie ahead, particularly as regards minerals exploitation and management. It may be that a new dispensation can be achieved which will allocate the resources of Antarctica on equitable principles, founded upon the concept of the 'common heritage of mankind'. However, if the Treaty is to be successfully

[1]Non-Aligned Movement: a large group of nations (101 members participated in the New Delhi conference) that actively refuses to be politically or militarily associated with either the Western or the Soviet blocs.

reformulated along these lines, there will have to be considerable 'real estate' concessions and surrenders of power by the original guardians of Antarctica. Unfortunately, however, most Treaty 'club' representatives believe it would be quite unrealistic to entertain the notion that sovereignty will be abandoned (Brennan 1983*a*); and it is considered unlikely that the consultative parties will make any significant concessions to the 'common heritage' approach (Auburn 1984). Thus, the heart of the problem relating to Antarctic resources—the absence of consensus as to the legal status of the territory—is likely to remain untouched. There is no agreement amongst the nations of the world as to who, if anybody, owns the resources of Antarctica or can purport to exercise jurisdiction over its vast land mass. To make the moral issue yet more complex, moreover, the Antarctic Treaty 'club' members believe that the majority of non-Antarctic Treaty nations at the UN discussions do not have the best interests of Antarctica at heart; and that these countries will therefore not adhere to voluntary restraints of the kind which the present Antarctic 'guardians' have nurtured for the needs and benefit of this continent. In the light of these complexities, progress in establishing a viable minerals regime for Antarctica is likely to be tortuous and protracted. Assuming that negotiations finally result in a new regime for the continent, what features is it likely to embody?

4.1.2. Possible models for a minerals regime

Several attempts have been made to predict the outcome of present deliberations and negotiations; and these give considerable insight into the range of possibilities for an Antarctic minerals regime (Alexander 1978; Pinto 1978; Mitchell and Tinker 1979; Lovering and Prescott 1979; Rich 1982; Mitchell 1982, 1983; US Department of State 1982; Auburn 1982; Quigg 1983; Vicuña 1983).

Whilst progressive thinkers propose complete internationalization or the trusteeship of Antarctica under United Nations administration (e.g. Anonymous 1978), **claimant** Treaty 'club' members appear—initially at least—to have favoured a co-operative arrangement, with responsibility for Antarctica vested in a condominium of these Treaty 'club' members (under a scheme analogous to that operative in the New Hebrides before its independence). Neither possibility seems a likely solution at present (see Mitchell 1982, 1983 for overview, although it is important to emphasize that there is now a broad measure of international consensus for the concept of internationalization of Antarctica under the umbrella of the 'common spaces' concept—Anonymous 1978). Other compromise suggestions, such as the recognition of 'diluted' or 'tempered' sovereignty (as in Svalbard; Mitchell 1982; Rich 1982; Sollie 1983), or a regime based on jurisdictional ambiguity, envisage an essentially cryptic continuation of the present *modus operandi* of Antarctic management. These suggestions are nebulous, however, and concomitantly difficult to evaluate. Mitchel (1982, 1983) believes a solution incorporating jurisdictional ambiguity is the most likely to be adopted. Indeed, the Treaty 'club' appears to be moving slowly in that direction, through pursuing the formulation of a minerals regime as a protocol to the Antarctic Treaty (*ECO* 1983, 1984; Tucker 1983*b*; *Nature*, 1983*c*, 1984; Auburn 1984).

The need to establish Antarctica as an 'international park' and to bring the exploration and exploitation of its resources under close international control, has also been debated in other fora, notably by internationally based conservations groups, such as The Friends of the Earth and Greenpeace. These have proposed that Antarctica and all seas and land south of the Antarctic convergence should become a 'Natural Wilderness Area and World Heritage' controlled by an international body composed of all interested states and non-governmental environmental organizations. Many of these conservationists totally oppose the exploitation of any natural resource within this region. Indeed they are against any activity which would initiate industrialization and endanger Antarctica's pristine environment, for this would conflict with their concept of the Antarctic wilderness as a World Heritage area (Lovering and Prescott 1979; Brewster 1982; Kimball 1983*b*). It is doubted that the environmentalists' proposals will ever be implemented (Mitchell 1983). Nevertheless, the serious messages with inter-generational consequences and responsibilities for all, voiced by environmental protection groups such as Friends of the Earth, the Sierra Club and, lately, Greenpeace (Brewster 1982; *ECO* 1983, 1984*a, b*), are of too fundamental a nature to be ignored and should be incorporated into any regime for future Antarctic mineral management. Whilst this has been recognized by the Treaty 'club', who have contracted out environmental investigations and protection procedures to SCAR scientists, it may legitimately be questioned whether this is sufficient: and whether environmental protection is too far-reaching in its implications to be left to scientists to handle alone, and especially only to those with Treaty 'club' connections.

The adequacy of existing environmental impact investigations and protection in Antarctica becomes even more suspect when it is compared with the investigations conducted, and legislation subsequently implemented, in the Canadian Arctic. Compared with these Canadian studies, the present 'in-house' environmental discussion and rules for Antarctica are highly theoretical and amateurish in practice (see also Auburn 1982, 1984; *ECO* 1984*a*, *b*). For example, it is generally recognized that environmental pollution at and around scientific stations in Antarctica is an embarrassment (Lipps 1978; Brewster 1982; personal observations of the author). Ironically, the US Antarctic program purports to comply with the Antarctic Treaty standards for pollution control (Antarctic Treaty Recommendations VIII—11 and the Code of conduct for Antarctic Expeditions and Station Activities) but there are no formal regulations (Todd 1983). Experience from some two decades of serious Arctic resources development has proved at least one point: the most environmentally sound, safe, and economic technologies result when technical and environmental information is non-proprietary, freely shared amongst all concerned and subjected to public and competitor comment or emulation (Roots 1983).

A number of the suggested 'solutions' for Antarctica emphasize the example which has now been provided (under UNCLOS III) for managing the minerals of the Deep Sea bed. Deep-sea bed mining is to be monitored through a central United Nations' management agency, the Deep Sea Bed Authority. The arrangements include a scheme designed to protect investments already made in pioneer deep-sea mining activities by national and international mining consortia (PIP, Preparatory Investment Protection). The Law of the Sea Convention has been signed by at least 130 nations and will officially come into force when ratified by a minimum of 60 nations. It is not known when this number will be reached (there are only nine ratifying nations up to the present), but it will probably not be before 1985 (Simmonds, personal communications 1983, 1984). Even after the Treaty comes into force, the mining and mineral management regime for the deep oceans which it incorporates could be challenged by an alternative 'Reciprocating States Regime' embodied within a mini-treaty, as proposed by the United States and under serious consideration in four other industrialized nations (United Kingdom, Federal Republic of Germany, France, and Italy; Brown 1983). Such a regime is permissible under international law (Kronmiller 1980); and the enactment of concomitant national 'interim' legislation to regulate deep sea-bed mining is being watched with great care by the international legal community (Simmonds 1983–). The motivation behind such legislation may only be purely precautionary: but its proliferation is making the majority of UN General Assembly member nations highly critical of what they see as an attempt—by the industrialized states—to fragment the 1982 Convention by selecting from amongst its provisions those which they perceive as serving their best interests. However, irrespective of whether this alternative scheme is ultimately implemented, it does at least seem that even a conservative mini-treaty could not fail to recognize the deep-sea bed as *res communis*,[2] nor to incorporate at least some of the emerging rule of customary international law regarding the 'common heritage of mankind' (Simmonds, personal communications 1983). Such elements might then make such a mini-treaty acceptable to nations which are committed to the enforcing of the common heritage principles into the realms of international law.

Since Antarctica is at present neither recognized as *res communis*, nor for its greater part even as *res nullius*, there is an inescapable difference between the judicial status of the deep-sea bed and Antarctica. A fundamental legal question, accordingly, is whether Antarctica could, in time, be changed in status to *res communis*. Anonymous (1978) and Auburn (1982) have argued convincingly that the present claims to

[2]According to this principle, the sea bed, the subsoil beyond the limits of national jurisdiction are held in common by all States; and sovereignty or sovereign rights may lawfully be claimed or exercised by any State over such areas, except in cases of waiver or prior-occupation. It thus follows that the resources of the deep sea bed may, at present, be appropriated by any State or private enterprise on a non-exclusive basis (Kronmiller 1980).

sovereignty over Antarctica are of doubtful validity in international law and could be successfully challenged before the international legal tribunals. If this were to happen, this would open up a greatly increased spectrum of possibilities for negotiation of a new Antarctic regime. However, of infinitely greater practical significance is the fact that almost all nations that are party, or have acceded to the Antarctic Treaty—both claimants and non-claimants—are committed, in principle, to the 1969 United Nations resolutions regarding the 'common heritage of mankind'. New Zealand has probably been the most vociferous of the claimant countries in its support for the 'common heritage' concept (Auburn 1982; Roberts 1983). Sentiments in favour of withdrawing a territorial claim in order to support the idea of including Antarctica as 'common heritage of mankind', to be administered by the UN, or some other international agency, have also recently been expressed in Norway (Sollie 1983).

Other, non-claimant states are also strongly in favour of this concept. Thus, India, which is a forceful proponent for the international legal recognition of common heritage, on the grounds that this principle is a necessary foundation for a new world economic order, has recently established itself as a new Antarctic power. It has built a sizeable scientific base on Antarctica which is to be permanently manned from 1985 (Fig. 1.1) and is conducting regular, scientifically orientated expeditions (*Nature* 1982, 1983*c*; Ali and Richardson 1983). India thus satisfied all the requirements needed to become a member of the inner circle of the Treaty 'club', and duly applied for consultative status in 1983, and was granted this a month later, in September of that year. Another member of the Non-aligned Movement and Group 77[3] nations (Argentina) is also a consultative member of the Antarctic Treaty System, whilst others have recently also made specific moves to involve themselves more seriously in Antarctic activities (Table 4.1). For example, Brazil, a Group 77 member, was admitted to the Treaty club as a consultatve member, also in September 1983 (an unprecedented move, since Brazil does not have the normally required scientific qualifications). The Peoples' Republic of China, which is to some extent in favour of United Nations involvement in Antarctica, has acquired Antarctic expeditionary experience, by courtesy of New Zealand and Australia, and has recently (1983) acceeded to the Treaty 'club'; (Kimbal 1983; *Nature* 1983*c*; Appendix A2). Furthermore, at the last (12th) consultative preparatory meeting in Canberra, Australia, acceding members of the Treaty were for the first time, granted observer status at Treaty 'club' meetings. All this means that there is a rapidly expanding forum from which to negotiate for an equitable 'common heritage' solution to Antarctica's mineral regime (Table 4.1)[4]. The stability of any future minerals management regime must, however, inevitably rest upon international consensus as to the implications of the 'common heritage' concept: and the Law of the Sea Conference has been invaluable in reaching towards and formulating the appropriate principles, which may now serve as a precedent for an Antarctic scheme.

The 'common heritage' concept reflects an emerging rule of customary law that all mankind must benefit from certain resources of the world, such as, for example, the mineral wealth of the deep-sea bed. However, both the definition of such benefits and the status of the legal obligation to share them remain ambiguous, and—at this point in time—precatory rather than imperative:

> However, there is not, at the present time, a specific legal obligation on States to share with the international community revenues derived from deep seabed mining. It is true that some States hold that contributions to the international community from deep seabed resource exploitation are legally obligatory. Nevertheless, a significant number of other States, among which are those having a vital interest in deep seabed mining, are of the view that the increase in the global availability of minerals constitutes a benefit to the world community sufficient to meet any kind of obligation arising out of the concept of the common heritage of mankind.
>
> Any contribution of revenues pursuant to domestic legislation would not necessarily constitute recognition of an international legal obligation. Revenue sharing provisions of pending United States legislation must be viewed in the context of the official United States position that the common heritage of mankind will be legally defined only by a future comprehensive Law of the Sea treaty.

[3]Group 77: a group of predominantly Third World nations (120 members in 1980), permanently represented at the United Nations as a caucusing bloc, to harmonize their negotiating position on matters of trade and development. Originally established by 77 nations in 1967 during UNCTAD I.

[4]*Footnote added in proof:*
Recently the Republic of Cuba has acceded to the Treaty; this has not been incorporated into Table 4.1.

> It must be borne in mind, as well, that States may reserve their rights against evolving customary international law. There is ample authority for the rule that a State which persistently objects to a developing rule of customary law retains its rights under pre-existing law. This proposition may be invoked by those States, including the United States, which insist that the common heritage of mankind will be legally defined and have legal effect only by a widely acceptable treaty establishing a new regime for the deep sea bed. There is no legal obligation imposed upon those states by virtue of the concept of the common heritage of mankind and will be none until they abandon their objection to the arguably emerging rule of customary law or consent to be bound by a new conventional regime embodying that concept.
>
> (Kronmiller 1980, pp. 516–7)

Clearly, there will be no firm standard against which to monitor progress of negotiations towards an Antarctic minerals regime, on an equitable basis within the framework of the 'common heritage' concept, until the Law of the Sea Convention is recognized by all. Furthermore, the 1982 Convention is arguably a triumph more for geography than for equity; for the coastal nations (and the archipelagic ones) have come out of the whole exercise very well indeed. All this does not augur well for a United Nations' General Assembly approach to Antarctica.

These then are the scenarios at present generally envisaged for an Antarctic minerals regime. However, there are any number of other possible approaches which could be applied to the challenges of crystallizing a code for the management of Antarctic minerals. Why have these apparently failed to spark widespread interest or excitement to generate global consensus for an optimum solution? The answer probably lies partly in the following advice from Jenks (1958): 'It is a weakness of all these approaches that they seek to solve the problems of Antarctica by transplanting ideas evolved elsewhere rather than by evolving a regime based on the problems and needs of Antarctica itself.' (p. 370.)

There is undoubted cogency in this observation. It thus suggests that the key to the difficulty may lie in taking the converse approach, and in focusing instead on the immense benefits that the Antarctic (mineral) resources have to offer in solving the problems, and providing the needs, of the global community. In other words, it may well be that better understanding of the significance of Antarctica's resources may provide a more concrete framework for the negotiation of an equitable minerals regime for the region. What, then, are the benefits offered by Antarctica: and how may these be used to meet the problems and needs of the world?

4.2. ANTARCTIC SCIENCE AND RESOURCES TO SUPPORT A NEW GLOBAL ORDER

4.2.1. Antarctic science: a stimulus towards awareness of global unity

In retrospect, the designers of Antarctic science projects were wise to capitalize on the major innovations of IGY in setting up targets for new programmes. It is worthwhile listing some of the important scientific gains over the last 25 years, to see whether any achievements that have been made may serve as foci for stimulating further awareness of the importance of Antarctica and its mineral resources. Here is a list of a few of these achievements:

1. Recognition that steady growth and decay of large ice-sheets is intimately tied to long-term variations in global climate.

2. Widespread utilization of remote sensing from aircraft and satellites yielding accurate information about the seasonal dynamics of the Antarctic sea-ice, the surface of the Antarctic ice-sheet and the shape of the underlying bedrock. This has allowed detailed estimations of the ice-sheet thickness, its volume, the driving stresses and flow lines within the Antarctic ice-sheet, and the outline of ice-drainage basins. Another important advance has been the recognition of major sub-glacial water bodies below the thick Antarctic ice-sheet.

3. With the recognition that the Antarctic ice-sheet may be subject to periodic surging and/or disintegration, great strides have been made towards documentating and understanding of whether such events are the driving forces behind glaciation and other major climatic changes. Large scale ice-sheet surges could, for example, increase the albedo of the southern oceans sufficiently to reduce world temperatures and start northern hemisphere glaciation. Smaller events could influence shorter term temperature fluctuations. It has been estimated that possible surges in the Antarctic ice-sheet could raise the global sea-level by as much as 15 to 20 m. Small-scale surges of the west Antarctic ice-sheet could have caused more frequent fluctuations in the sea-level of 2–3 m.

4. Ascertainment that the apparent stability of our current climate is due in part to the response of the polar heat sink to variations in energy input elsewhere. To understand climate, including man-induced variations thereof, it is accordingly crucial to understand the heat-balance at the polar regions, the processes which can modify it, and the response times related to them.

5. Recognition that the global weather machine is very sensitive to events in the polar regions and that recent temperature changes may have been three- to four-fold larger at the polar regions than elsewhere.

6. Demonstration by chemists that oxygen isotope ratios fossilized in the ice-sheets provide an accurate chronological record of global climate changes, that nitrate concentrations may provide information on solar activity, and that other chemicals (e.g. DDT) and particles collected from carefully dated ice-cores provide unequivocal data on man-induced, as opposed to natural (e.g. extra-terrestrial or volcanic), pollution of the atmosphere, and the fall-out history of nuclear explosions. Ice-core studies have also shown that dramatic changes in atmospheric carbon dioxide content occurred at the close of a glacial period, as well as during relative warm intraglacial periods, over short time spans of less than 1000 years. Carbon isotope records from these cores may provide the key to a better understanding of these climatic shifts.

7. Greater understanding of atmospheric exchange and mixing between Antarctic air and air masses from the mid-latitudes through widespread utilization of data from air-chemistry and physics over a wide range of earth and environmental sciences.

8. Vast improvement in the understanding of the fundamental composition of the Antarctic lithosphere, and the recognition that the Transantarctic mountains divide the continent into two major sub-regions of profoundly different crustal age and architecture.

9. Immense gains in the understanding of the formation of the southern oceans and the Scotia Arc, as well as the shape and subsequent break-up of history of Gondwanaland, of which Antarctica formed a central part. This has yielded valuable insight into the formation of Antarctica's passive continental margins and their associated thick sedimentary basins, and has brought together information relevant for estimating the fossil fuel potential within these sediment traps.

10. Demonstration by marine earth-scientists that only after the final isolation of Antarctica from is surrounding continents did a truly Antarctic circumpolar current develop which had a major impact on global ocean water circulation and climate.

11. Recognition that the cold nutrient-rich Antarctic waters account for more than half of the bottom waters of the world's oceans.

12. Demonstration by marine biologists that these polar waters are some of the most biologically productive regions in the world and contain some of the earth's greatest source of high-value protein in the form of krill, and that these waters buffer one of the simplest known ecosystems through the direct food-chain between the immense whale and the tiny shrimp-like krill.

13. Recognition that the Antarctic region is crucial for the study of radio-communications and solar terrestrial physics. Charged particles emitted during solar flares are responsible for magnetic storms, cause auroras (plasma phenomena) and give insights into the electrical coupling between the upper and lower atmosphere. An important source for the earth's electrical field is the interaction of the solar wind with the magnetosphere, and a unified view of the electrodynamic system of the atmosphere requires studies of energy transfer processes in high latitude magnetic field-line regions. For practical purposes it has been demonstrated that these energy transfers are of fundamental importance towards understanding thunderstorms, lightning, and electrical power interference in transmission lines, which can cause major power blackouts.

(References related to these summaries can be found under subject listings in *Current Antarctic Literature and Antarctic Bibliography* 1951–1983).

There is a unifying factor in all these Antarctic scientific gains: they greatly increase understanding and awareness of the important role Antarctica plays in the system Planet Earth; and of the scientific keys which Antarctica holds for monitoring the state of health of this system. For example, scientists concerned with the global climatic warming through the growing carbon-dioxide (CO_2) content in the atmosphere, owing to ever-increasing burning of fossil fuels, have implicitly asked, Will anthropogenic changes in the Earth's atmospheric CO_2 ultimately force a dramatic change in world climate? Will the CO_2-

induced warming of our planet, which will occur during the next hundred or so years, activate such a change? Carbon dioxide measurements on Antarctic ice-cores have so far captured the interest of those interested in climates of the past; they must now surely also capture the imagination and concern of those involved with global climates to come (Broecker 1984).

One of the major advances in the natural sciences over the last decades has been the emergence of an awareness of the planet as a total ecosystem with all the socio-economic implications this entails. The awe-inspiring view of the earth from the Apollo spacecraft, irreversibly caught the imagination of both physical and socio-economic scientists, catalysing a new appreciation of the need to understand the structure and dynamics of global interrelationships (Boulding 1983). In the life-sciences for example, ecosystem approaches have flourished, encouraging better modelling of processes related to destruction and conservation of finite, exhaustible resource systems. Indeed, it was this type of approach which in part led to the international agreements to curb the decline of the whale population in Antarctic waters, and helped to establish the new multidisciplinary BIOMASS program (Biological Investigations of Marine Antarctic Systems and Stocks). Geologists have since similarly improved their understanding of the workings of the planet by integrating physical manifestations at the earth's surface from widely separate areas into a unified, interrelated network of earth-reactions; from this emerged the concept of Earth Sciences in the late 'sixties and early 'seventies. Many former Departments of Geology have changed their nomenclature to Departments of Earth Sciences, and have simultaneously incorporated subjects such as environmental studies and resources evaluation. Today, Antarctica can no longer be excluded from these exciting developments. In the economic sciences, too, open and closed system analyses at different scales have vastly improved understanding of the interdependencies of the Earth's different national and international economic systems. The time may soon come when Antarctica will also have to be included in these dynamic economic models.

4.2.2. Antarctic resources for global equity and needs

As awareness of global interdependence has increased; as interest in systems theory and methodologies has grown; as apprehension over resource-depletion and environmental deterioration has mounted; and as socio-economic stability has proved ever more elusive, so there has been a dramatic shift in the general social conception and definition of 'need'. Instead of the satisfaction of need being regarded as a matter of charity (and concomitantly suspect), a new notion of distributive justice, based on legitimate rights and expectations, has evolved (e.g. Ward and Dubois 1972; Meadows *et al.* 1972; Tinberger 1976; Schachter 1977; Bosson and Varon 1977; OECD 1979; Smith 1979; Barnet 1980; Brandt 1980, 1983). In this way, 'need' has become a measuring standard for equity or real equality. Schachter (1977) recognizes this as a profound change in the perception of values, one with significant implications for international policies:

> By treating need as a standard of equity (and of 'real equality'), we reduce considerably the vagueness and indeterminancy of the concept of equitable sharing. Need, after all, can be ascertained in some objective way; there is even a widespread consensus on basic requirements of human beings everywhere.
>
> (Schachter 1977, p. 10)

The equation of 'equity' with 'need' provides the opportunity for the quantification of equity, making it amenable to numerical analysis and more practical solutions. This is no mean achievement, even though defining needs on an international scale remains a complex issue:

> (Nor) is it enough to calculate, as many have done, the essential needs of individual human beings for nutrition, health, shelter, education, security, and other factors. The more important and complicated issues involve the definition of needs in the light of social processes and their interaction. For example, in determining the food requirements in a country one can begin with reasonably determinative facts such as nutritional requirements, population estimates and available food supplies. But then to define needs in a prospective and operational sense, it is essential to consider the interpretation of several factors that determine supply and demand: the changing means of production, the input of capital and technology, the incentives to produce, the system of tenure and redistribution, the market and credit arrangements, prices, export demands and several other conditioning elements. These factors are dynamic, and through their complex interactions the level of need for the people of a given region is determined. Obviously, then, the criterion of need does not provide simple answers for a single problem such as food (resources). It does, however, set a standard that can in principle be applied by objective factual inquiry.
>
> (Schachter 1977, pp. 10 and 11)

Since the principle of equity has been given a measure of practical application in collective decision making, the implications of this new response for the norms and structures of the international system are beginning to emerge. The standards of minimal global needs are being hammered out in the international fora, through declarations, resolutions and so forth, increasingly cast in terms of human rights and obligations. Through the expectations thus engendered, these statements provide the catalyst for the formulation of new legal rules. For much of the 'sixties and 'seventies the world was 'super-saturated' with concepts of equality and sharing; new economic order and with the need to improve social stability, optimize the use of resources, control environmental pollution and so on. In the 'eighties it appears that practical solutions are beginning to crystallize, and growth to meet these 'utopian' objectives is now gaining momentum (e.g. Calder 1983).

What is the significance of these developments for Antarctic mineral resources? The Law of the Sea (or the Moon Treaty) negotiations have great importance in this regard, for–although they cannot be taken as binding precedents for a new Antarctic regime—**they nevertheless set a firm trend of 'linking demand based on need to entitlement, derived from accepted principles'** (Schachter 1977; see also Anonymous 1978). For example, it is now widely accepted that access to the sea and to a share of the deep ocean mineral resources is due to all peoples or states under the principle of common heritage. Similarly, if it can be shown (as argued in Chapter 3) that there is indeed a need for platinum as a 'cleaning agent' to curb and reverse increasing deterioration of an undisputed common heritage—the atmosphere of the earth—there is surely a powerful argument for the need to appropriate this 'cleaning agent' as a common heritage as well. The contention becomes even more cogent, taking into account the reality that the pollution in question is predominantly caused by those who are—at present—able to control the supply of the 'cleaning agent', and who, *a priori*, have little economic incentive to adopt appropriate 'cleaning' processes. This provides a powerful argument and rationale for the internationalization of Antarctic resources. Crucial to any impetus towards global control, however, is the *extent* of the needs of the world community: and whether these are sufficient to justify both commencement of mining in Antarctica and the placing of such operations under international control. The first of these issues raises, at the outset, the question of the gains that may be expected from Antarctic mining: and this, in turn, depends upon the value of the region's mineral deposits. Such 'value' is the function of many different factors, however, and cannot be entirely objectively gauged—as emphasized by Dixon (1979):

> 'Because the needs of mankind change from time to time and place to place it is never possible to define a set of geological objects called mineral deposits in purely geological terms. The phrase 'value to mankind' needs some explanation . . . [However,] whatever the economic or political structure of the world in which a mineral deposit is worked, there will always be some criteria by which the effort put into exploiting it may be compared with incentive for doing so.'
>
> (pp. 5 and 6)

In the immediately foreseeable future there is no likelihood that land-based resources outside Antarctica will be exhausted, thus giving rise to real shortages. The main stimulus for the pursuit of a 'common-heritage of mankind' minerals regime for Antarctic resources lies, therefore, in the opportunity this may provide for the development of alternative supply sources to meet the global needs for social, political, and economic justice and security, as well as to minimize the risks of developing instabilities and inequity, pending the establishment of a new socio-economic order. The proposed regime to control sea-floor mining of metalliferous nodules, for example, represents a giant step in this direction and holds out great promise for optimizing social welfare and providing such support for a new world order. It is, indeed, a significant enough step for the prestigious *Mining Journal* (1982, p. 318) to advocate the signing of the Law of the Sea Convention; between the mineral riches of the Deep Seas *and* Antarctica, all major mineral needs can be met, and in sufficient quantities to influence world markets.

Bearing in mind the destabilizing effect on the world economy of disruptions in the supply of crucial raw materials, an important question to consider then is the strategic and social values of the mineral resources of Antarctica. How important are these to global needs?

4.2.3. Strategic and social values of Antarctic resources in a destabilized world

It is noteworthy that the problem of ensuring long term availability of mineral resources has reached alarming dimensions over the last decade. This development may be attributed, *inter alia*, to the following factors:

1. Social, political, and economic upheavals in Third World countries richly endowed in mineral resources, and the consequent re-evaluation of the wealth of their possessions (i.e. copper in Chile, Zambia, and Zaire; tin in Bolivia, Malaysia, and Indonesia; cobalt in Zambia, Zaire, and Cuba, bauxite in Surinam, etc.).
2. Increasing economic and political risk by those industrial countries largely dependant on the import of minerals from the Third World countries (i.e. the OECD countries).
3. Unstable frameworks of cooperation between different governments, and between many governments and multinational industries.
4. Uneven global distribution of raw materials and rapid increases in the price of energy resources, especially oil.
5. Fast deterioration of the environment in several key areas of the globe.

The significance of the first three of these factors is best appreciated in the light of the political instability resulting from decolonization, especially in Africa. Vale summarizes this as follows:

> While fears of nationalization and political instability, particularly in the 1970's, further inhibited the operations of First World mining corporations in the Third World, the cost of exploiting new mineral resources has also impinged on the availability of raw materials. A number of problems arise from the increasing costs attached to the development of new mining ventures. At the centre lies the same vexatious problems associated with the uncertain relations between mining corporations and the Third World host countries which have increased mining lead-times with the result that greater investment is on risk for a longer period and an accompanying economic price has to be paid. Moreover, the recurrent economic problems of the Seventies—high rates of inflation and low growth rates—have meant that all other things being equal, the costs involved in both new mining and infrastructural development have been affected by cost-push inflation. Given the nature of the profit factor in the *modus operandi* of the mining multinational corporations, they are reluctant to commit large sums of investment to projects with long and uncertain lead-times in geographic regions with high political volatility.
>
> (Vale 1981, pp. 172–3)

As a result of these realities, investment in exploration and exploitation in the Third World by multinational mining corporations has significantly declined (Tinbergen 1976; OECD 1979; Brandt 1980, 1983).

The practical result of these political developments in the 1960s and 1970s was to heighten the awareness of the industrialized nations of the vital importance of access to raw materials. Their resultant rivalry for Third World mineral wealth has led to increasing competition and instability in the world economic system and, arguably, to increasing metastable political and economic subordination of these Third World countries. The anxiety thus generated is clearly reflected in the *Interfutures* OECD report (1979), which warns as follows:

> In most cases, the countries which possess the reserves or which produce the materials are not the centres of consumption. This is true for both EEC and Japan and, in the case of certain raw materials, for the United States also. For the OECD countries the most crucial situations are those where there is, at the same time, a high regional concentration of reserves or production and high dependancy on supplies from Eastern countries, from developing countries or from South Africa.
>
> Apart from political uncertainties in general, and direct or indirect export restrictions in particular, a problem of concern to some countries, notably the EEC countries and Japan, may be the considerable influence of multinational mining companies in the relevant mineral markets. Even if there is no actual indication that these companies intend to discriminate against specific countries, the question remains: what may happen if there are prolonged shortages for political reasons? Markets with only very few suppliers are naturally more liable to breed discrimination than those where there are a good number of competitors.
>
> Moreover, the tensions between materials availability and political or social constraints related to environmental quality or even broader social concerns will increase still further in the future. Hence there may be a proliferation of public procedures for approving new investments together with a lengthening of the time taken to obtain authorisations. In many cases such authorisations will be based on *purely political considerations rather than on a cost/benefit estimate.* Admittedly in cases of severe shortage, political attitudes may change rapidly, but to bring new mines into production, or to reactivate mines or processing plants after they have been closed down, takes many years.

Over the last decade, events in the Middle East have underscored this disturbing dependence of many of the world's nations on foreign suppliers of crucial raw materials. The resultant socio-economic vulnerability of these nations to uncontrollable interruptions in the supply of these materials, such as oil, has also been starkly highlighted. This in turn has provided a major driving force (impetus) towards rapid development of deep ocean mining of nickel- and cobalt-bearing manganese nodules: for the possibility of obtaining an assured and steady supply of raw materials, immune from disruptive political, economic and social factors, has assumed increased significance.

The uneven distribution of resources throughout the world makes it possible for supplier states to discriminate against certain countries, particularly on social, cultural, and political grounds. This leads in general, to anti-social global pricing inimical to the best interests of the world as a whole (Beenstock 1983). As a result, concern over the unbalanced geographic distribution of mineral reserves is a recurring theme in most reports regarding the physical availability of raw materials for future decades. The situation is as follows for the 21 most important minerals covered in Table (4.2):

1. Forty-four per cent of reserves are held by the industrialized countries (OECD plus South Africa), 23 per cent by the eastern countries and 33 per cent by the developing countries.
2. Almost 90 per cent of the reserves held by the industrialised countries are in the United States, Canada, Australia, and South Africa.
3. The U.S.S.R. possesses more than 80 per cent of the reserves of the socialist countries, but it must not be forgotten that prospecting in China is still in its earliest stages.
4. In the developing countries, too, many of the reserves are held by a small number of countries: Brazil (25 per cent), Chile (9 per cent), Indonesia (7 per cent), Zaire, Papua New Guinea, and India (4 per cent).

It is thus evident that Western Europe and Japan, most of the East European countries, and about 70 per cent of all developing countries have only very limited mineral reserves.

Concerning seven of these commodities (chromium, columbium, manganese, molybdenum, vanadium, platinum, and asbestos) more than three-quarters of the measured and indicated reserves occur in three countries only. There are 15 minerals in relation to which more than 75 per cent of reserves are found in a mere five countries, and, of the 21 minerals cited, there are only two (copper and zinc) in respect of which these five countries' share is less than 60 per cent (OECD 1979).

The effect of this skewed mineral distribution on the power of supplier states to exercise socio-economic and political leverage is probably best illustrated by the case of South Africa. The significance of South Africa as a source of many of the world's key minerals has been discussed by many minerals experts (van Rensburg and Pretorius 1977; OECD 1979; Pretorius 1979; Vale 1980; U.S. Senate Committee on Foreign Relations 1980; House of Lords 1982; Neethling 1983; see Andor 1985 for overview). Tables 4.3 and 4.4 list 22 vital minerals, and rank South Africa's role in Western world supply in terms of exports, production, and reserves. In terms of exports it will be seen that South Africa leads in the supply of no fewer than 7 of the minerals listed, and holds a dominant position in the export of at least eleven others. To this end, therefore, South Africa is a minerals 'supermarket'. Such a rich endowment has stimulated maverick behaviour within South Africa: and there are at least two examples of blatant socio-economic exploitation **rooted in this mineral wealth.**

a) Economic exploitation in the form of global market control and monopoly pricing. This is probably best illustrated by the diamond industry, which is almost exclusively controlled by the Anglo-American Group, through the De Beers' Central Selling Organization (CSO). Today, more than 80 per cent of the world's rough diamonds are marketed through, or in co-operation with CSO (Thompson 1982). Sales by CSO in 1983, at \$ 1599 million, yielded profits of more than \$ 420 million, whilst the book value of their diamond stock increased to \$ 1852 million (Oppenheimer 1984).

b) Social exploitation of human resources in the form of black labour mobilization and control. This is probably best illustrated by the unrivalled importance of gold mining to the South African economy and its dominant role in sustaining the apartheid system. In 1980, gold production accounted for 60 per cent of the output of the country's mineral mining industry and contributed almost 20 per cent of its GDP. Overseas sales represented some 33 per cent of total exports and financed well over 50 per cent of South Africa's imports. Gold mining is, and always has been, the principal engine of South Africa's overall economic activity and growth. It is the backbone around which much of the country's industrial modernization has developed, and it is the base on which the extensive superstructure of white minority rule and privilege

TABLE 4.2. *Twenty-one industrial raw materials with their geographical distribution which emphasizes the importance of those areas which were part of Gondwanaland*

Raw material	Share of three countries 1977	Share of five countries 1977	Regional distribution measured and indicated reserves (1977); country and percentage share
Basic metals			
Iron	59.4	76.6	USSR (30.2) **Brazil (17.5)** Canada (11.7) **Australia (11.5) India (5.8)**
Copper	44.9	58.7	USA (18.4) **Chile (18.5)** USSR (7.9) **Peru (7.0)** Canada (6.8) **Zambia (6.4)**
Lead	47.8	61.4	USA (20.8) **Australia (13.8)** USSR (13.2) Canada (9.5) **South Africa (4.1)**
Tin	50.2	68.1	**Indonesia (23.6)** China (14.8) **Thailand (11.8) Bolivia (9.7) Malaysia (8.2)** USSR (6.1) **Brazil (5.9)**
Zinc	45.8	58.6	Canada (18.7) USA (14.5) **Australia (12.6)** USSR (7.3) Ireland (5.5)
Light metals			
Aluminium	62.8	74.8	Guinea (33.9) **Australia (18.6) Brazil (10.3)** Jamaica (6.2) **India (5.8) Guiana (4.1) Cameroon (4.1)**
Titanium	59.0	74.1	**Brazil (26.3) India (17.5)** Canada (15.2) **South Africa (8.6) Australia (6.6)** Norway (6.4) USA (6.0)
Alloying metals			
Chromite	96.9	97.9	**South Africa (74.1) Zimbabwe (22.2)** USSR (0.6) Finland (0.6) **India (0.4) Brazil (0.3) Madagascar (0.3)**
Cobalt	63.0	83.5	**Zaire (30.3) New Caledonia (18.8)** USSR (13.9) Philippines (12.8) **Zambia (7.7)** Cuba (7.3)
Columbium (Niobium)	88.5	95.3	**Brazil (76.6)** USSR (6.4) Canada (5.5) **Zaire (3.8) Uganda (3.0) Niger (3.0)**
Manganese	90.5	97.7	**South Africa (45.0)** USSR (37.5) **Australia (8.0) Gabon (5.0) Brazil (2.2)**
Molybdenum	74.3	86.9	USA (38.4) **Chile (27.8)** Canada (8.1) USSR (6.6) China (6.0)
Nickel	54.5	76.8	**New Caledonia (25.0)** Canada (16.0) USSR (13.5) **Indonesia (13.0) Australia (9.3)** Philippines (9.0)
Tantalum	72.7	84.8	**Zaire (55.0) Nigeria (11.0)** USSR (2.9) North Korea (6.4) USA (6.1)
Tungsten	69.6	80.6	China (46.9) Canada (12.1) USSR (10.6) North Korea (5.6) USA (5.4) **Australia (2.7)**
Vanadium	94.9	97.2	USSR (74.8) **South Africa (18.7) Chile (1.4) Australia (1.4) Venezuela (0.9) India (0.9)**
Accessory metals			
Bismuth	47.9	60.9	**Australia (20.7) Bolivia (16.3)** USA (10.9) Canada (6.5) Mexico (6.5) **Peru (5.4)**
Mercury	65.2	78.3	Spain (38.4) USSR (18.2) Yugoslavia (8.6) USA (8.6) China (4.5) Mexico (4.5) Turkey (4.5) Italy (4.1)
Precious metals			
Silver	54.9	76.5	USSR (26.2) USA (24.8) Mexico (13.9) Canada (11.6) **Peru (10.0)**
Platinum	99.5	99.9	**South Africa (82.3)** USSR (15.6) Canada (1.6) **Columbia (0.3)** USA (0.1)
Other			
Asbestos	81.3	91.8	Canada (42.7) USSR (32.3) **South Africa (6.3) Zimbabwe (6.3)** USA (4.2)

Countries in bold print were once part of Gondwanaland.
Modified from Michalski (1978), and OECD (1979).

TABLE 4.3. *South Africa's mineral reserves and sales relative to the rest of the world*

	Reserves				Exports			
	Western world		World		Western world		World	
Mineral commodity	Rank	%	Rank	%	Rank	%	Rank	%
Manganese (ore)	1	93	1	78	1	43	1	33
Platinum group metals (600m depth)	1	91	1	77	1	92	1	74
Vanadium (metal, 30m depth)	1	91	1	49	1	86	1	55
Chrome ore (300m depth)	1	75	1	73	1	75	1	40
Gold	1	64	1	51	1	68	1	59
Fluorspar	1	45	1	34	2	27	3	20
Andalusite/sillimanite	1	45	1	34	1	58	n/a	n/a
Vermiculite	2	29	2	28	1	81	1	78
Diamonds	2	23	2	21	2	25	n/a	n/a
Uranium (metal, up to \$50 Lb U_3O_8)	2	14	n/a	n/a	3	24	n/a	n/a
Antimony	2	18	3	5	6	9	7	6
Asbestos (fibre)	4	8	4	5	2	15	3	11
Zirconium (metal)	2	12	2	10	2	23	n/a	n/a
Coal	3	13	6	10	3	17	5	12
Zinc (metal)	4	6	5	5	17	1	n/a	1
Lead (metal/concentrate)	4	5	5	4	7	10	11	10
Nickel (metal, 600m depth)	5	8	7	6	6	12	8	7
Titanium (metal)	3	17	4	15	n/a	n/a	n/a	n/a
Silver (metal)	5	6	6	4	10	1	14	1
Cobalt (metal)	6	3	7	2	n/a	n/a	n/a	n/a
Iron ore (30m depth)	6	6	7	3	6	12	8	7
Copper (metal)	8	2	10	1	6	4	7	8
Tin (metal)	12	1	14	1	9	3	10	3

Source: Chamber of Mines of South Africa (1980); Neethling (1983). n/a: not available.

rests. South Africa's gold mining industry has been decisive in shaping many of the country's social and political institutions and policies, and its system of migrant labour mobilization, organization, and control (see for example van Onselen 1982, and Innes 1984). It has been argued that in this respect the apartheid system is more than a system of institutionalized racism (i.e. the denial, by statute and regulation, of basic human rights to sections of the population on the grounds of race and colour). Rather, the roots of the apartheid system lie deep in the organization and structure of the gold mining industry, because the profitable exploitation of low-grade gold-bearing rock has required, at each stage of the development of the mining industry, the massive application of cheap labour, referred to as 'low-grade human energy' (AAM 1981).

In 1982, the mining industry absorbed well over 722 000 workers, 86 per cent of whom were black and 68 per cent of whom were employed on the gold mines (SAIRR 1983). The white: black wage ratio at these mines stood at 5.5:1; the net operating surplus (pre-tax profits) amounted to 189 per cent (194 per cent in 1983) of the industry's aggregate wage and salary bill (S.A. Minerals Bureau, private communication 1984). It is the exploitation of cheap black labour which makes the South African gold and other extractive industries amongst the most profitable in the world today.

For several reasons, the extent and consequences of socio-economic injustices thus outlined pose questions of serious concern for the world community. For example, in a recent study, Vale (1980) develops the hypothesis that the latent racial crisis in South Africa will fracture the nations of the Atlantic Community in the minerals arena. Vale argues that such fracture will **result** from the uneven spread of economic ties, **reflecting unequal mineral utilization and access**, between different nations and South Africa.

TABLE 4.4. *Mineral commodity imports from South Africa by the United States, Japan and countries of Western Europe (1979)*

Commodity	United States	Japan	United Kingdom	West Germany	France	Italy	EEC
Andulasite group							
Minerals	90	71	65	15	—	32	n/a
Antimony	6	—	90	15	7	—	10
Asbestos	4	34	18	12	8	49	12
Chrome ore	48	52	79	66	24	23	50
Coal	49	30	1	19	37	16	23
Copper	—	9	6	10	—	3	7
Diamonds	52	7	n/a	26	3	7	n/a
Ferrochrome	79	71	18	54	37	33	49
Ferromanganese	45	—	46	4	—	18	25
Fluorspar	23	27	—	12	—	—	7
Gold	—	—	66	5	16	59	56
Iron ore	—	5	10	8	7	7	6
Manganese ore	4	15	46	73	41	53	45
Manganese metal	31	—	74	52	54	73	78
Nickel	4	18	—	17	14	23	7
Platinum group							
Metals	55	28	58	9	17	52	28
Vanadium	87	90	—	6	20	—	25
Zinc	4	—	—	9	—	—	3
Zirconium	1	7	7	16	—	1	4

Expressed as a percentage of total imports of that commodity.
Source: Neethling (1983).

These factors have profound significance for mineral exploitation in Antarctica, especially if, as is happening with the Law of the Sea, such exploitation can be placed under international management (under the auspices of an international agency linked to the United Nations). Under such guardianship, the exploitation of specific mineral commodities could be controlled to best advantage—not only to counteract supply discrimination—but also to support the combating of other malpractices which militate against the general social welfare of the global community.

Environmental considerations and public attitudes to resource development will, in future, also have a great impact on the global or regional availability of raw materials. New societal hostility towards resource development stems from the detrimental impact of mineral extraction and processing on land, on water, on the atmosphere and on the social environment. Today, therefore, social values place important constraints on the definition and dynamics of mineral reserves. Many communities do not wish to have minerals developed within or near their boundaries, because of adverse environmental and social side-effects. In a certain number of advanced countries, the prevailing attitude is increasingly one of letting such developments occur 'elsewhere' (OECD 1979).

In summary, whilst it is impossible to predict the eventual outcome of the potential confrontation engendered by the imbalanced distribution of minerals in the earth's crust, what does appear certain is that the major immediate consequence of skewed mineral supplies will be higher inflation rates and further socio-economic disruption on a global scale. To counteract this, it is essential to develop a stable new world economic order, in which dependence on unreliable mineral resource supplies is eliminated as far as possible. If the mineral reserves from the deep ocean basins **and** from Antarctica can be placed under international management on a 'common heritage' principle, this will provide an appropriate framework for the development of such a new order; the new mineral wealth of these regions will then be available as a buffer against socio-economic and political volatility and instability. Moreover, a steady supply of important minerals from these sources will help to prevent unacceptable anti-social behaviour (of the kind manifested in South Africa). If these goals are to be attained, however, delicate management and political restraint will be essential. This raises the question whether such restraint can be secured outside the 'common heritage' concept; and the last section of this chapter is devoted to discussion of this complex issue.

4.3. MAJOR ANTARCTIC NEEDS TO ENSURE GLOBAL GAINS

Should a minerals management regime be adopted otherwise than on a 'common heritage' basis, the risk will be extreme–especially in view of the location of certain potential mineral deposits in Antarctica, within areas over which rights of sovereignty have been claimed by differing states. The most significant potential endowments (as presently assessed) are those within the Dufek complex and (the oil resources of) the Weddell Sea. Both of these areas lie within parts of Antarctica claimed by Chile, Argentina, and the United Kingdom (Fig. 4.1). The exploitation of these deposits could therefore put the Antarctic sovereignty issue to severe test and undermine the most delicate pivot of the present *modus operandi* of the Antarctic Treaty. Recently ugly confrontations between these nations over sub-Antarctic sovereignty issues, specifically between Argentina and the United Kingdom (i.e. on South Georgia during the Falklands/Malvinas conflict), have heightened fears for further conflict in the Antarctic region. Should such a clash occur, this would have serious consequences far outweighing any benefits arising from future mineral exploitation in Antarctica. Accordingly, it is imperative that a minerals regime for Antarctica should *not* be based on the present rules of the Treaty. The fact that the Dufek complex—at present one of the few feasible areas for exploitation—is situated in an area claimed by no less than three nations, must demonstrate to the Treaty signatories the importance of resolving the sovereignty issue: and should encourage them to use all efforts to negotiate a sound equitable agreement. Furthermore, there is a growing commitment to implement the urgent global social reforms which are necessary to maintain world stability: and these are now, arguably, too significant to be left our of account in formulating a new Antarctic minerals regime. Poverty, hunger, unemployment, pollution, and other human rights violations are on the increase, and entail such grave projected consequences if allowed to continue unchecked that the need to reverse these ills must surely pose an ever-mounting challenge to the present chauvinism of the Antarctic 'elite'. The challenge may, however, generate its own dangers for continuing peaceful co-operation in Antarctica—for the attempt to meet it may shatter the present delicately balanced accord. It thus needs little further emphasis that it is vital—to preclude competition and conflict and to promote the well-being of all mankind—that a stable regime be formulated for Antarctica: and that this be place squarely within the framework of the 'common heritage' concept.

Antarctica, like the deep oceans, has untold potential to help provide the needed cures for current social conflicts. Accordingly, the important question regarding Antarctica and its resources is not the extent to which the 'profit' factor may be expected to induce minerals-exploitation. It is common sense that Antarctic mining profits are likely to be lower than those obtainable elsewhere; Antarctica's economic rent is lower simply by virtue of its inhospitable climate. The pertinence of Antarctic mineral wealth lies, rather, in the question whether its exploitation offers the chance of global gains beyond pure economic profits. To pose one possibility, is it feasible that by **merely increasing our knowledge of Antarctica's mineral wealth**, optimal resource utilization in other areas (with a concomitant increase in social wealth) could be stimulated? Economic theory indicates that this is so. Could Antarctica's potential resources also buffer social and political tensions outside its immediate region? There is no *a priori* reason to reject this possibility, especially in relation to platinum, with its now well established essential role in environmental pollution control. Environmental considerations cut both ways, however, and it is accordingly essential to examine the other side of the coin as well, and to assess the likely impact of the extraction of this (and other) mineral(s) on the fragile Antarctic ecosystem.

To date, scientific reports conclude that mineral exploitation developments, namely oil-drilling and mining for metallic ores, are both very local in their effects and that it is unlikely that exploitation could be on such a scale as to pose any significant threat to the terrestrial ecosystem in Antarctica as a whole (EAMREA 1979, AEIMEE 1981, 1983). The analysis in these reports is, however, theoretical and lacks practical or experimental support. Given these shortcomings, the available studies do not—and should not—satisfy those who, on environmental grounds, oppose resource development within and around Antarctica.

The environmental costs, risks, and returns of exploitation must be much more carefully considered. The Consultative Parties have not, in the past, been very forthcoming in funding such SCAR investigations, nor is there evidence that there will be much enthusiasm for enforcing environmental safeguards in the projected minerals regime (Auburn 1984). This does not augur well for the survival of a pristine Antarctic environment. Antarctica **needs** very careful and sophisticated scientific environmental monitoring and modelling, coupled with ongoing socio-economic evaluation of the risks and benefits of exploitation. Intuitively, one aspect of mining in an internationalized Antarctica seems certain: monitoring and controlling its effect on the local and global biosphere would be considerably easier, and perhaps even more cost-effective, than that of mining ventures under a variety of national flags. Again, this serves to emphasize the vital importance of placing Antarctica under international control, on the equitable basis of the 'common heritage' concept.

A pragmatic evaluation of the environmental risks entailed in further economic development—within the framework of a new global order—is provided in the *Interfutures* report of the OECD (1979).

> Is it a fact that, because of the indirect—and often cumulative—consequences of the scale of his economic activities, man is compromising the survival of the other animal and vegetable species and his own? This is a question to which no one can remain indifferent but it is necessary to distinguish clearly between the problem it poses—that of protecting humanity against the considerable and very uncertain risk inherent in destructive action on the ecosphere—and the more tactical issues like maintenance of the local environment or the elimination, at least partial, of pollution . . .(p. 56)
>
> Even though mankind must concern itself increasingly with the impact of economic activities on the environment in all its forms, the economic growth of the countries of the world taken as a whole can continue during the next half-century without encountering long-term physical limits. However, as we have seen, it may well be necessary to improve, in certain respects and above all on a national scale, the contents of that growth . . .(p. 61)
>
> Thus, protection of the physical environment does not for the moment constitute an obstacle to the development of economic activities. In a number of cases, it can even make for a more harmonious pattern of growth that better reflects the aspirations of the population. This is a field in which governments, both of developed and of developing countries, must be ever ready to act, whether it be to finance research to improve the state of our knowledge, or to take conservation measures, or to solve local difficulties which can be considerable. (p. 61)

There is no doubt that the environmental issues are going to have a major influence on any decision concerning exploration–exploitation in Antarctica, as they continue to do in the Arctic. In the latter region, the protection of the environment, in accordance with some politically or socially acceptable level, has emerged as the decisive factor in allowing otherwise economically profitable ventures to proceed (Roots 1983). Environmental concern and the protective rules ultimately adopted for Antarctica must be even more stringent and must be strictly enforceable, if only because of its natural uniqueness[5]. Moreover, it is apparent that the exploitation of Antarctica may have very wide-ranging consequences.

For example, the environmental effects of mineral extraction on Antarctic living resources must be carefully assessed because of the importance of these as potential world food reserves. The possible damaging effects on local and world climate will also be in the forefront of environmental debate and concern. The melting of the ice caps could have catastrophic results on the major cities and densely populated areas of the world: for, to take an extreme example, nearly all of them would be totally flooded by the concomitant 60-m rise in sea-level (Fastook and Hughes 1982). Although there is considerable scientific doubt about the likelihood of such an man-induced catastrophe, there is a greater consensus of opinion that smaller shifts in the global ice–water balance may, in the long term, be triggered off by shifts in the present climatic equilibria as a result of man-induced changes in the composition of the oceans and atmospheres

[5] *Footnote added in proofs:*
Recently the conservationists have called for the establishment of an international Environmental Protection Agency (AEPA; Greenpeace 1984). They envisage that such an agency would have primary responsibility not to a small group of governments, as is presently the case, but to the preservation of the Antarctic environment for the international community as a whole.

(Fastook and Hughes 1982). Present scientific models of the atmosphere show that projected global temperature increases are the result of steadily rising CO_2 levels in the atmosphere due to man's increasing fossil fuel burning. A small temperature increase might be sufficient to trigger, for example, the collapse of the West Antarctic ice sheet, resulting in a subsequent global rise in sea-level of from 5–7 m.

In summary, mankind faces an awesome task in balancing its need for earth resources against its need for a stable undamaged environment. In this study it has been argued that Antarctic resources can fulfil an active, and preferably a passive role in such a task, provided they can be controlled within a minerals regime based on a 'common heritage' framework.

In 1977, Pinto advised an important first practical step in the formulation of a final Antarctic minerals regime: the gathering of clearer and more comprehensive data regarding the resources and economic potential of the region. The Treaty nations, however, have not yet come close to meeting Pinto's challenge nor has any Treaty signatory been persuaded to fund such a study. How can this be explained; and how could such a comprehensive review best be conducted? The next chapter attempts to answer these questions.

5 ESTABLISHING AN ANTARCTIC MINERAL RESOURCE INVENTORY

In the early stages when it might have been politically possible to enforce sensible controls, the essential scientific advice was lacking. In the latter stages, when sound scientific advice became available, it was too late.

(The story of Antarctic whaling; B. Roberts 1977)

Those who cannot remember the past are condemned to repeat it.

(G. Santayana)

5.1. ANTARCTIC LOGISTICS AND MINERAL EXPLORATION

Despite many reported mineral occurrences, to date no mineral reserves have been shown to exist in Antarctica and only the documentation of resources of coal and iron has been seriously attempted (Splettstoesser 1980, 1983; Ravich, Fedoror, and Tarntin 1982; Rowley and Pride 1982; Rowley *et al.* 1983; Berhendt 1983). Most Antarctic geologists are content to accept that this is largely related to the lack of solid rock exposure (~2 per cent; Wright and Williams 1974; Rowley and Pride 1982; Rowley *et al.* 1983). Conventional prospecting wisdom, however, would probably support the notion that lack of discoveries of economic ore deposits might be better correlated with Antarctica's inaccessibility and related logistic problems, such as are vividly described in the accounts of explorer-scientists such as Cook (1777), Amundsen (1912), Scott (1913), Shackleton (1919); Ralling (1983), Fuchs and Hillary (1959). Even today, scientific endeavours are faced with formidable logistic and economic challenges. For example, exploratory geological investigations within an operating radius of 185 km in the Ellsworth Mountains (79°5′S 85°58′W; Fig. 1.1) from 3 December 1979 to 12 January 1980, required the establishment of a base camp for the scientist and technicians a month ahead of the start of the investigations. In real terms, this entire operation involved 78 Hercules (LC-130) flights and a total of 670 flight hours (at a cost of about $ 1600 per flight hour) to establish, maintain and break up the camp, while three UN-1N turbine helicopters flew 352 h in direct support of scientific work at the camp. Approximately 40 people remained at the camp for 60 days (Splettstoesser 1982, personal communications; Splettstoesser *et al.* 1982; Guthridge, Manager of US Polar Information Programme, personal communication 1983).

In the United States Antarctic Research Programme (USARP), there are about three support staff to every one scientist, and out of a total estimated budget of US $ 66 million for the 1982 programme, only 5.6 per cent was spent directly on scientific work, with less than 1 per cent available for the earth sciences (USARP 1982; see also Fig. 1.4). Since the United States Antarctic budget and logistic support outweighs those of other treaty nations (for example, the total 1982 annual budget for Antarctic research in the United Kingdom was $ 10.05 million; British Antarctic Survey 1982; Fifield 1982; *New Scientist* 1983; See Fig. 1.4.), the overall economic input for geological research in Antarctica is insufficient in terms of a realistic mineral exploration budget. Furthermore, during major geological expeditions, which last only a few months each year, rock outcrops receive only minimal examinations. Elsewhere in the world, key rock-exposures are visited by numerous geologists time and time again, whilst those in the Antarctic are rarely visited more than once, often receiving only scant attention. It is important to stress that it is inadequate to assess the resource potential of Antarctica on such scanty reconnaissance coverage. It is commonly assumed by academic Antarctic geologists that if studies in the exposed parts of Antarctica have failed to reveal mineral concentrations rich enough to be classified as ore deposits, they probably do not exist, or they will not be found. This philosophy may be as instrumental towards an under-estimation of the resource potential of the exposed parts as problems of potential over-estimation in the ice-covered area.

To illustrate the above, one might examine a case history of an established mining district in the southern Kalahari desert of South Africa, known as Bushmanland. Here, extensive bare-rock exposures outcrop sporadically, projecting like islands out of the vase expanse of sand, not unlike nunataks out of the Antarctic ice-sheet. Both terrains therefore offer similar excellent, though limited, rock exposures. Accessibility throughout the Bushmanland area, facilitated by a reasonable infrastructure, is infinitely better than in Antarctica. At one locality, Aggeneys, a road passes by some extensive rock exposures, which in several places display some of the world's most classic indicators of *in situ* mineralization. These indicators, known as gossans, are exposed at the surface, enabling geologists to see them from great distances away. Although copper occurrences have been examined here by numerous geologists and experienced prospectors between 1929 and 1970, the full base-metal potential of the area was not realized until 1971. Today, its reserves rank amongst some of the largest in the world. In 1980, one of three main deposits (Broken Hill) was brought into production at an initial milling rate of 11.25 kiloton per year, with an average grade of extractable reserve of 0.4 per cent copper, 6.35 per cent lead, 2.87 per cent zinc and 82.25 g/ton of silver, which will be sufficient for a life of about 30 years at this rate of production. In later years, attention will be focused on other deposits, discovered nearby, where it is already known that reserves exceeding 81.6 million tons exist at medium grades and 101 million tons at lower grades (Ryan 1982). This example underscores the unpredictable nature of the art of prospecting and the often lengthy lead-time needed to locate and evaluate, in retrospect, even the most obvious signs of an economic mineral deposit. Successful prospecting requires not only talent, knowledge, experience, and intuition that only few geologists possess, but also perseverance and economic investment over a lengthy data-collecting period, for which thus far little opportunity has been created in Antarctica.

Two further points serve to illustrate the difficulties in assessing the mineral potential of Antarctica purely on an exploration basis. Firstly, the success of prospecting has been vastly improved over the last decade through the increasing application of sophisticated theoretical models of geological environments in which ore deposits are believed to form, together with more accurate prospecting tools to locate them. Probably the most impressive recent example of this is the discovery of the Roxby Downs copper–uranium deposits in South Australia, near Adelaide, in 1975. Today one of the deposits, the Olympic Dam, is one of the world's major undeveloped mineral resources, and preliminary estimates indicate that it contains at least 2 billion tons of ore with an average grade of about 1.6 per cent copper, 0.6 kg/ton uranium oxide and 0.6 g/ton gold (Olympic Dam Project 1982). The unusual nature of this find is that the presence of mineralization was not predictable from any surface expressions. In fact, the economic ores occur at depths ranging between 350 and 650 m, beneath a hard rock 'blanket' of a younger age than those in which the ore minerals are contained. Western Mining Corporation started exploration of the Upper Proterozoic Stuart Shelf in this region, on the basis of:

(a) an ore genesis model, relating strata-bound copper deposits to alteration of basalts in basins;
(b) a promising tectonic setting of a rift, a major lineament and a basement high;
(c) a similar age and geology to that of the Zambian Copperbelt; and
(d) known copper mineralization at Mount Gunson, 100 km to the south.

Thus, the deposit was located through target drilling on a geologically conceived model constrained by some deep geophysical anomalies (Trueman and Maiden, personal communications, 1983 and 1984 respectively). As such, this discovery is unprecedented and heralds a new era of prospecting for ores at deeper levels in the crust than has hitherto been believed possible. Furthermore, the fact that the deposit discovered was in Middle Proterozoic rocks, that the geological model turned out to be at least partly wrong, and that the geophysical (gravity) data had been misinterpreted, continues to underline the need for a large 'luck factor' in mineral exploration strategy.

Thus, and with specific relevance to Antarctica, lack of indicators of mineralization at the surface of rock exposures is no longer a good measure of the mineral reserve potential in a specific area.

Secondly, indicators of near-surface mineralization are today relatively easy to prospect for, using, for example, chemical tracers and geophysical signatures. Lack of consistent exposures does not eliminate discoveries of major deposits, using such methods; the gold and nickel deposits of south-west Australia, and the copper deposits below the Kalahari sediments in Botswana and Namibia, occur in extensive terrains that can boast even less exposure than in many parts of Antarctica. However, in these parts of Australia and southern Africa, the cover 'blanketing' the signs of mineralization is thin and, in the Australian case, autochtonous (i.e. *in situ*). It is much more difficult to trace chemical anomalies to mineralized sources where such an overburden is thick and allochtonous (i.e. *out of place*), as for example in areas covered by thick glacial deposits (tills) and still more so where such a blanket is in motion as in the case of desert sands. Nevertheless, beneath the former, such prospecting has been succesfully carried out, for example in Ireland, during base-metal prospecting in the mid sixties to mid-seventies. Below the latter, such as in the Kalahari desert of Botswana, diamond bearing kimberlite pipes of small diameter can now be located where an overburden of more than 100m of sand hides their existence.

In principle, there is no reason to believe that modified approaches cannot be successfully adapted to trace ore deposits beneath the thick moving ice overburden of Antarctica, although this will require a planned and systematic approach from which meaningful results cannot be expected for a considerable period of time. We will return later to this challenging prospect. At present, however, it is more pertinent to focus on the question of which methods are at our immediate disposal to obtain a reasonable quantitive estimate of Antarctica's mineral potential. It appears that such an assessment can only be attempted using scientific hypotheses, integrated with knowledge of rock formations within Antarctica and ore deposits external to this continent.

5.2. ANTARCTIC RESOURCES STUDIED IN THE FRAMEWORK OF GEOLOGICAL AND TECTONIC-METAL ANOMALY MAPS OF GONDWANALAND

It is now generally well accepted that large mineral deposits must underly the Antarctic ice-sheet. The logic of this belief rests on the fact that such deposits occur on every continent, and since the basic geological architecture of the Antarctic is known to be similar, there is no *à priori* reason to believe that equivalent deposits do not also occur on this continent. This seems to be a valid first-order assumption. Indeed, as remarked upon by Skinner (1979), analysis of historical observations on the growth and decline of the mining industries in different regions of the world indicates that equal volumes of the earth's crust, when assessed on continental-size units, contain approximately equal amount of non-ferrous metal wealth. Thus, factual statements that Antarctica is about equal in size to that of the USA and Mexico together, or to Canada and Greenland combined, or about one and a half times the size of Australia, speak for themselves *vis à vis* the mineral resource potential of Antartica. But can we do better than state such a 'sweeping' resource assessment? Even more rewarding, can we equate an estimate based on the above arguments with a more reliable disaggregated method of resource accountancy?

Some 180 million years ago, the continents of the southern hemisphere formed a single land mass or supercontinent known as Gondwanaland,[6] in which Antarctica occupied a central position (Fig. 5.1.). During the disruption of Gondwanaland, continental fragments were successively separated from Antarctica at various times, starting with Africa and South America (about 150–160 million years ago), India (120 million years ago), Australia (80–100 million years ago), and New Zealand (about 70–80 million years ago; Fig. 5.1, inset). The precise kinematics of separation is still subject to serious scientific investigations, as is the original configuration of smaller continental fragments of West Antarctica, such as the Antarctic Peninsula, in this Gondwanaland framework. On a broad scale, however, agreement on the fit of this supercontinent is remarkably consistent (Du Toit 1937; Smith and Hallam 1970; Smith, Hurley, and Briden 1981; de Wit *et al.* 1985; Fig. 5.1). Moreover, it is known that this Gondwanaland had been in existence for at least 500 to 650 million years, Thus, an argument that related mineral deposits which formed during the

[6]The concept of Gondwanaland as a supercontinent was first introduced in 1883 by the Austrian geologist Suess. Gondwana, or land of the Gonds, was derived after an ancient aboriginal tribe who are believed to have inhabited a large part of central India. In this region, geologists in 1872 first used this name for a system of rocks which contained characteristic freshwater plant and animal fossils, later found to be present on all southern hemisphere continents. Today, the term Gondwana is accepted as the more correct term for the name of the supercontinent, although in this book its original nomenclature is maintained.

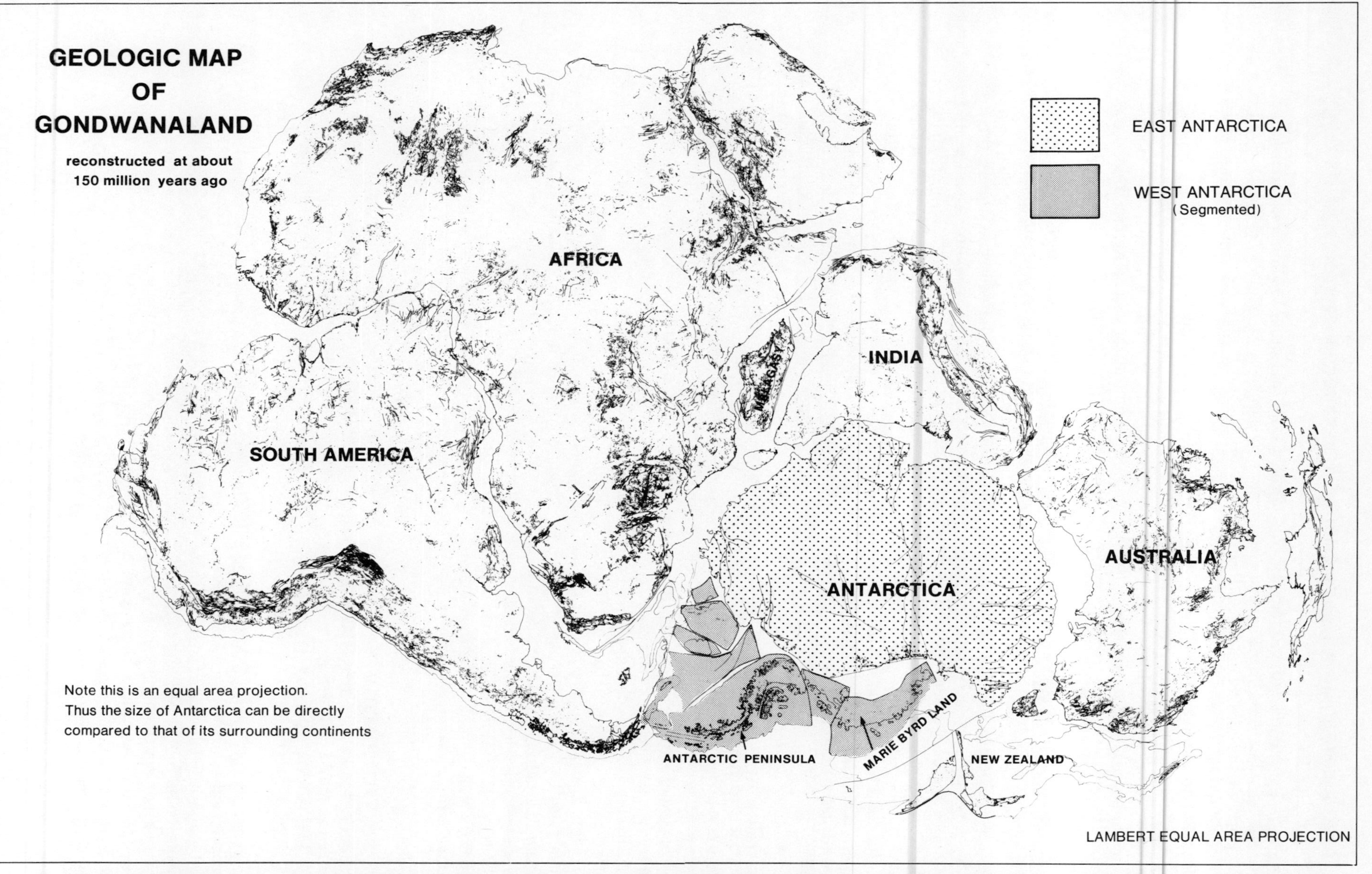

Fig. 5.1. Geologic map of the southern continents, reunited into the supercontinent Gondwanaland, as it existed about 180 million years ago. In this Gondwanaland context, large parts of East Antarctica can be directly compared to the surrounding mineral enriched regions in southern Africa, India, and Australia. West Antarctica was known to be segmented into smaller fragments (micro-continents) during the dispersal history of Gondwanaland. Most of these micro-continents have close geological affiliations with the Andes and other parts of South America. The Antarctic Peninsula for instance, is a direct geologic extension of the Andean mountain chain. Detailed knowledge of the history of Gondwanaland, its geology and its ore deposits can thus be used in a systematic way to predict the mineral potential of Antarctica. (After de Wit *et al.* 1985.)

TABLE 5.1. *Expected number of deposits to be discovered in exposed areas of Antarctica, on the basis of an assumed success rate of 1:100 for deposits once identified*

Type of deposit	Andean sector	Ellsworth zone	Ross zone	Antarctic shield
Ferrous metals	0.025	0.01	0.006	0.0029
Base metals	0.075	0.007	0.006	0.0015
Precious metals	0.035	0.001	0.003	0.0014
Other	0.025	0.002	0.005	0.0015

Source: Wright and Williams (1974).

period from about 800 to 150 million years ago, by whatever ore-producing process, could be present throughout the entire Gondwanaland area, also seems perfectly acceptable, provided the rock records in the now widely separated areas are compatible.

It is on this second-order assumption that several attempts have been made to predict and evaluate the mineral potential of Antarctica at various levels of sophistication (Du Toit 1937; Termor 1951, Fairbridge 1952, Schnellman 1955; Anonymous 1956; Chalmers 1957; Runnels 1970; Wright and Williams 1974; Ericksen 1976; Quartino and Rinaldi 1976; Elliot 1976; Kanehira 1979; Lovering and Prescott 1979; Zumberge 1979*a;* Borman and Weber (1983). In the most detailed quantitive study, Wright and Williams (1974) calculated a density of known mineral occurrences per unit area within the continents surrounding Antarctica, and extrapolated this data statistically onto Antarctica, using sub-regions with a similar geographical history. From this they formulated a resource potential of Antarctica in terms of expected numbers of deposits to be discovered in each of the sub-regions (Table 5.1).

Unfortunately, this work does not discriminate qualitatively between mineral deposits, and because the raw data used in the calculations is not provided, it cannot be judged on a serious basis, since statistically 'apples and oranges' are being compared. Furthermore, the work does not allow for mineral deposits that have not yet been discovered outside Antarctica. Thus, in the light of major new deposits, such as Broken Hill in South Africa, or the Roxby Downs in Australia, as previously described, as well as many other sizeable finds throughout the Gondwanaland fragments since the early 'seventies' that are not incorporated in this study, the results must be considerably under-estimated. More realistic statistical approaches such as these developed by French and Canadian economic geologists (Pelissonier and Michel 1972; Economic Geology Division, Canada 1980) and that, for example, used to evaluate the mineral potential of a vast unexposed region such as the Sahara Desert (Allais 1957), as well as methods tested on a smaller scale by predominantly Canadian scientists in active mining areas of North America (Agterberg, Chung, Divi, Fade, and Fabbri 1981), must be incorporated if an evaluation of Antarctica's mineral resource potential is to have a more reliable foundation. It must be stressed at this point, however, that the statistical treatment and interpretation of mineral resource data is neither universally accepted nor adequately understood (Skinner 1976, 1979; De Young 1981; Zwartendyk 1981, Harris 1984).

A more serious scientific flaw in the Gondwanaland-structured studies so far is the exploitation of the underlying assumption of a single configuration of Gondwanaland throughout geological time, from the earliest preserved rock sequences to about 800 million years ago. Estimates of Antarctica's mineral potential in the Gondwanaland framework have, for example, included sub-regions of significant aerial extent, which formed and stabilized between about 3800 and 2500 million years ago (the Archaean shields or cratons, Fig. 5.2), and which contain some of the greatest riches of their respective continents.

The configuration of Gondwanaland that far back in time, if it existed at all, is not known. It is an issue of current debate amongst earth-scientists, but it appears that the existence of this super continent, in its presently known form, before about 800 million years ago is probably erroneous. Extrapolations of ore deposits formed prior to about 800 million years ago must, therefore, be attempted with utmost care, and in unison with other geological and palaeomagnetic constraints.

This is of particular importance when assessing specific mineral resources such as the

Fig. 5.2. Gondwanaland reunited, showing the disposition of some major geological (tectono-chronologic) subdivisions, discussed in the text. At present, the concept of Gondwanaland to estimate the mineral potential of Antarctica can only be confidently used since the time of the Late Proterozoic (600 to 800 million years ago). The configuration of Gondwanaland prior to that time is not known with the same degree of certainty. Most of its resource potential must therefore be attempted using a different strategy, as discussed in the text. For clarity, early- to mid-Proterozoic regions and belts (2500–1000 million years) are not outlined. Rocks of this age are known to extend from at least the 'southern boundary' of the Archaean cratons to the very edge of Gondwanaland such as in Peru (Arequipa inlier) and Australia (Georgetown and Coen inliers). These rocks form a sialic basement to later mobile belts and are known to have been rejuvenated in many placed. Such a process is another complex variable to be incorporated into the resource evaluation of Antarctica. (After de Wit *et al.* 1985.)

precious metals, chromium, nickel, manganese or others with which these old cratons (Fig. 5.2) are disproportionately endowed relative to the total known resources of the world, but which, amongst these regions themselves, are not apparently homogeneously distributed. Some of the reasons for the latter observation are well understood. For example, gold has been concentrated in extraordinary quantities at the uppermost levels of some of these stable crustal remnants, (e.g. the Kaapvaal and Zimbabwean cratons of southern Africa), whilst precious gemstones are more abundant at the surface levels of the southern Indian craton and its age equivalent in parts of the Antarctic craton (Grew and Manton 1979; Matsueda, Motoyoshi, and Matsumoto 1983; Fig. 5.2). Yet it is clear from known similarities of the rock records preserved in these old cratons throughout the world that they underwent similar histories of formation. Thus, other geological reasoning must be dovetailed into a more realistic comparison between these cratons. To this end, it can be established from complementary geological information that these 'similar' cratons, or parts thereof, are uplifted and eroded to expose different crustal depths at the present-day surface levels (Fig. 2.2).

With such knowledge, a further order of reasonable assumptions can be formulated to evaluated mineral resources in provinces or linear belts with rocks spanning a similar age and with a comparable geological history of erosion, sedimentation, alterations by heat and distortion (internal deformation) as well as additon of new material from the mantle below the crust. For example, whilst the history of geological processes in the Andes from northern to southern Chile is almost identical, this continuous belt has been differentially uplifted and eroded to expose a deeper copper-poor section in the south, compared to a shallower, copper-rich section in the north (Figs 2.2 and 5.2). Similarly, it is now known that the northern extremity of the Antarctic Peninsula is in parts highly uplifted relative to its southern extension (Dalziel 1983). Thus, in an inverse way, knowledge of Andean geology and its mineral occurrences, integrated with Antarctic geological observations, can be extrapolated to predict with greater confidence the presence or absence of, for example, porphyry copper deposits and their supergene enrichment zones along sections of the Pacific margin of West Antarctica.

Ideally, one needs to approach these inherent problems of resource evaluation within specific geological provinces as an exercise in multivariant resource analysis (c.f. Singer and Overshine 1979; Agterberg 1981; Singer and Mosier 1981; Harris 1984). This is a major challenge because the original extents and three-dimensionality of geological provinces are often speculative. In many places these provinces have been partially to wholly 're-digested' during subsequent periods of thermal and mechanical distortions associated with orogenesis (periods of mountan building), and/or are 'blanketed' over by thick covers of new sedimentary layers. The effects and extent of the former process, namely crustal recycling, are contentious issues under careful scrutiny today by earth-scientists from many parts of the world. Quantification of recycling has, in fact, been part of a long-standing and seemingly irresolvable debate, but recent work, especially in the field of geochemistry, using isotopic signatures as tracers, indicates that this process is now sufficiently well understood to be of real value in differentiating between regurgitated crustal material, which may have contained previous ore deposits, and the new ore-forming elements added from the mantle below (Kay 1980, 1985; Watson 1980; Karig and Kay 1981; Cohen and O'Nions 1982*a* and *b;* DePaolo 1984; Hofmann 1984). The above considerations have great bearing on evaluating polymineralized terrains, since the overprinting process may either disseminate or further concentrate previous crustal mineral resources, in addition to further enriching the entire terrain with new resources from below.

Figure 5.2 encapsulates some of the spirit of this problem. This diagram shows that some of the earth's oldest known rocks (≧3.5 billion years) are present at the centre (i.e. the southern African and Indian cratons) as well as along the eastern and western 'edges' of Gondwanaland (Pilbara craton, NW Australia and the Guyanan or Amasonian shield, South America, respectively). Additionally, there are large areas of Precambrian rocks, older than 2 billion years, exposed at surface level along the southern 'edge' of Gondwanaland (such as the Ariquapa Massif, southern Peru; the Georgetown and Coen inliers, NE Australia) as well as close to Gondwanaland's northern margin (i.e. the Requibat Rise, NW Africa; the Tuareg shield, north central Africa; the Uweinat inlier of Sudan, Libya and Egypt, NE Africa; and northern India.

Thus it appears that more than 90 per cent of

Gondwanaland about 150 million years ago ($0.8–1 \times 10^8$ km^2) may have had an underlay of Pre-cambrian crust or lithosphere formed during the first half of the earth's history. It is also known that this early lithosphere was already disproportionately endowed with many valuable 'incompatible' elements, as reflected by its profound chemical heterogeneities in the form, for example, of rich metalliferous ore deposits (see for example Watson 1978).

Figure 5.2 also indicates that up to one third of the southern parts of this 'Gondwanaland lithospheric cap' may have been repeatedly overprinted by three mobile belts or major periods of lithospheric reactivation, since the time when Gondwanaland may have attained its approximated configuration. These belts are:

(i) the Late Proterozoic to early Palaeozoic mobile belt, often referred to as an Pan-African in Africa and Brazilian in South America;

(ii) the Palaeozoic mobile belt of the Samfrau geosyncline of du Toit (1937), herein referred to as the Gondwandes; and

(iii) the Mesozoic–Cenozoic Andean mobile belt.

Thus, in this southern section of Gondwanaland, a quantitative approach towards understanding polycyclic mineralization and redistribution of metallic 'incompatible' elements related to each of these overprinting periods must be an important aspect of resource evaluation for large tracts of the Gondwanaland lithosphere, and by inference, Antarctica. For example, it may well be that high grade copper (or gold) deposits originally formed in older, more internal parts of Gondwanaland have been chemically redistributed to form part of the many lower grade deposits within the external Andean mobile belts of the now separated fragments of Gondwanaland.

Similarly, the covering of well defined geological provinces by new sedimentary blankets at the surface is a major variable to be incorporated in a regional resource evaluation. Along the southern margin of Gondwanaland, a thick 'blanket' of young volcanic and sedimentary rocks (Fig. 5.3) covers earlier mineral-rich rock provinces to the east of the Andes in an irregular pattern. Such 'blankets', or remnants thereof, are known throughout the entire recorded history of the earth. They may hide mineral resources in the rocks below, or may themselves contain significant sedimentary mineral concentrations derived from sub-surface processes of erosion and re-working of these underlying 'rock-foundation'. Thus again, the concept of crustal exposure level is extremely important in evaluating the mineral resource potential over large areas. The old cratonic areas and slightly younger provinces of Antarctica are known to be overlain in places by at least one such 'blanket', known as the Beacon Supergroup. Unfortuantely, the extent of this cover is not known and has to be extrapolated, using remote sensing, placing extra constraints on a resource evaluation of Antarctica.

Finally, it is important in any worthwhile resource assessment to incorporate some computable degree of real social value. In economic terms, for example, the distinction between a mineral resource and a mineral reserve is simple. The former grades upward in rank to the latter, if it complies with an economic standard defined by a particular socially acceptable norm. This 'bottom-line' is set by a required rate of return on investment when exploiting the mineral deposit at a given point in time. Accessibility, infrastructure, mining technology, and mineral beneficiation are all variables incorporated in this standard, as are sizes, shapes, and concentrations of each deposit. Thus mineral resources assessment becomes a dynamic issue when set in an economic and socio-political framework. Ore deposits of equal total mineral contents come in different sizes and concentrations. They may be of very large volumes (tonnage) but with very low concentration (grade), or they may be small deposits of high grade. Variations of these properties are fundamental in defining a social value of mineral resources and should therefore be incorporated into resource evaluation, such as that of Antarctica, despite the fact that statistically the natural proportionality between these variable is far from understood (Skinner 1979).

To conclude, we have set ourselves the difficult task of an Antarctic mineral resource accountancy which can be subjected to rigorous auditing. Is this an attainable goal? The results of a recently initiated project to evaluate Antarctica's Mineral Resource Inventory, given the inherent constraints, have been encouraging to the extent of stimulating efforts towards constructing Gondwanaland, and thus Antarctic, tectonic-metal anomaly maps (de Wit and Bergh 1984–5). Three major steps are involved in the production of these maps:

1. Calculating specific elemental concentrations at point sources (mines, deposits, mineral

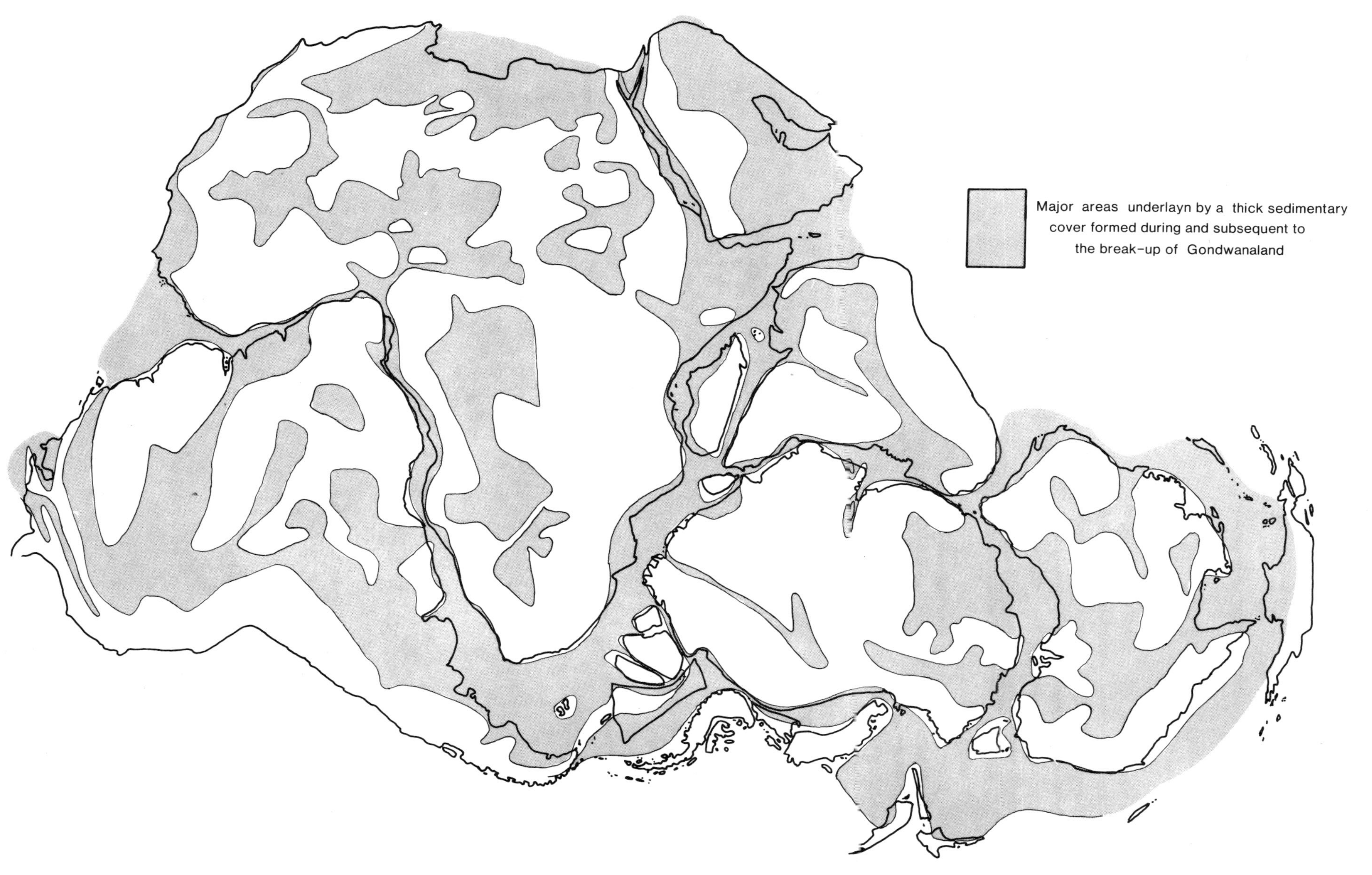

Fig. 5.3. Gondwanaland reunited, showing the extent of major areas of thick sedimentary 'blankets' formed during and since the continents dispersed. Such 'blankets' increase the uncertainties in a mineral-resource evaluation of Antarctic rock sequences which formed prior to the Gondwanaland dispersal history. (After de Wit *et al.* 1985.)

occurences) in the now separated continental fragments of the Gondwanaland lithosphere, using known and estimated tonnages and grades from existing and historical mines.

2. Normalizing this data.

3. Computer rotation and transfer of the data onto a Lambert equal area tectonic base map of Gondwanaland, and contouring of the resultant patterns.

Digital integration of this map with the geochronological and geological data from the same computer programmes recently developed for the geological map of Gondwanaland (de Wit *et al.* 1985) will then allow a more accurate estimate of the mineral resource potential of predetermined sectors of Antarctica, and of Gondwanaland in general.

It should be relatively simple to test the results of this work by equating it with a theoretical estimate of resource assessment in an average crustal volume of a continental size similar to Antarctica's.

Ultimately, however, the real challenge lies in a more applied test. Can the strategy for such a test be realistically formulated today? The answer is almost definitely 'yes', as revealed in Fig. 5.4. This diagram shows that the thick ice-

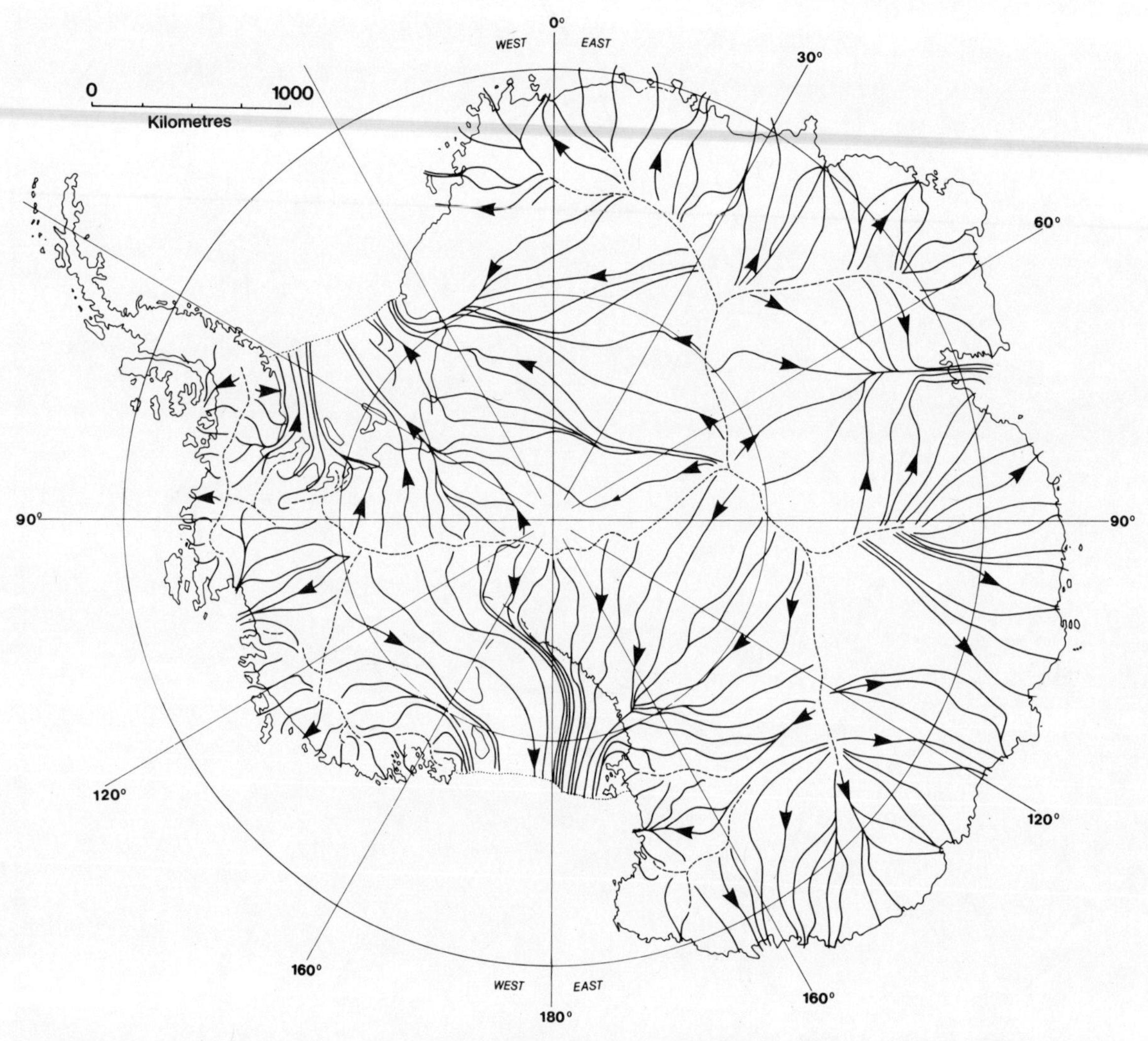

Fig. 5.4. Major flow directions of the Antarctic ice-sheets (modified after Drewry 1983). Because the rock layers of Antarctica have been systematically sampled by erosion of these ice-flows, sediment dispersal patterns of resource-index minerals around Antarctica can reveal much about the sub-ice mineral potential of this continent. Scientific investigations of this nature may provide a future test for more theoretical resource evaluations of Antarctica.

sheet covering the Antarctic continent moves in an established, almost radial pattern away from the central Antarctic topographic plateau and well defined ice-drainage basins. This 'ice-river' has been slowly eroding the underlying rock surface by means of mechanical scraping and grinding, carrying the resultant rock particles and flour to the surrounding oceans, where this debris has been dumped for millions of years. In a systematic way, rock layers of Antarctica have been sampled by this 'infinite frozen-river'. Therefore, any mineral deposit or manifestation thereof must now be represented in the sediment samples of glacial moraines and marine deposits on the continental shelves and ocean floors around Antarctica, and radially outward from there in a simple pattern related to particle size, density, and ocean water circulation. Recent marine sediment coring off the Antarctic Coast to the north-east of the Weddell Sea off Queen Maud Land, for example, has consistently revealed the presence of gold particles (Anderson *et al.* 1982 and personal communication 1983).

Systematic sampling of these sediments by marine coring thus offers the most accurate and practical method presently available to assess the near-surface mineralization potential of this continent. With the growing knowledge of the flow patterns of both the ocean waters and the Antarctic ice, mineral deposits might eventually be accurately traced to within smaller definable areas. A similar approach to search for the inland source area of recycled fossil pollen found in sediment cores off the Ross Ice Shelf has been successfully attempted (Truswell 1983 and personal communications), and in an inverse way such a technique has yielded new finds of dense meteorite concentrations on Antarctic glaciers. Given such smaller target areas, exploration drilling through the ice might become a tantalizing prospecting tool when combined with local detailed geophysical probing through the ice cover.

This futuristic exploration of Antarctica's mineral wealth may yet lie several decades into the future, but there is no question that one day the Antarctic earth-science community must confront their particular responsibility in this regard and face the challenge of attaining an accurate resource inventory of this continent. If they wish Antarctic 'scientific colonialism' to be projected in a favourable light by future generations, the 1980s may be the most timely period to initiate a concerted international study of Antarctica's mineral resources. Serious consideration must be given to the establishment of an International Geological Survey to carry out this type of pioneering work systematically. Harmful political decisions for Antarctica may well ensue in the absence of better factual data of this nature.

6 EPILOGUE

Antarctica, ice-bound, remote, rich in minerals, ripe for mining—albeit at great cost. What is its significance to modern-day man?

This study has attempted to explore and explain its vast potential. It is not, however, easy to draw the threads together and point to a definite conclusion as to the appropriate use of this great continent. There are so many seeming contradictions in assessing the extent of its mineral wealth, the feasibility of mining in its cold climate, the cost of its exploration and exploitation. And, to make matters yet more complex, it is clear that these technical issues cannot be divorced from larger considerations stemming from Antarctica's unique position in the world. Antarctic affairs are, today, tied into world affairs, scientifically, technically, economically, politically: they have become interlinked to world social issues and particularly those related to the environment. Technical answers alone do not suffice. Socio-political factors—on a global scale—must also be drawn into account: for there is increasing understanding, especially among scientists, that the continent cannot be viewed in isolation, but must be seen as part of a global ecosystem in which each part is intrinsically and inextricably linked to the larger whole.

Against this background—and all that has already been canvassed in this study—it remains to spotlight the key factors bearing on the future of Antarctica. First among these is the Antarctic Treaty. At present, the continent, in terms of the Treaty, falls under the jurisdiction of twelve founder nations, together with four subsequently adopted members. These sixteen have managed Antarctica as a type of trust territory for the general benefit of mankind; and there is a general consensus that they have done their job well. In 1991, however, the Treaty 'comes of age' and the question as to whether or not there is a need to continue the *modus operandi* of the Antarctic Treaty, with very few concessions to the rest of the world, is now being diligently examined. The founder parties to the Antarctic Treaty are adamant that they should retain their guardianship and that an international body should be kept out of Antarctic government. The Antarctic 'foster parents' are determined to 'hang on to their baby': partly because they mistrust a larger, global management; partly because of the sovereignty issue (for no less than seven out of the twelve founder parties have laid permanent claim to large parts of the continent).

Diametrically opposed to the founder group are an increasing number of nations which have no part in the present Treaty system and who contend (with much force and considerable legitimacy) that Antarctica forms part of the common heritage of mankind. Today, they regard Antarctica as sufficiently 'grown up' to be placed under the jurisdiction of a more widely representative international body, such as the United Nations. Indeed, this group has been instrumental in having Antarctica placed for open discussion on the annual United Nations General Assembly's agenda since 1983.

Midway between these two opposing groups is a third unit consisting of those nations which, through scientific and technological leverage, have won increasing involvement in Antarctic affairs, albeit without voting power. This group has expanded rapidly (from a single nation in 1961 to more than fifteen countries in 1984); and all indications are that it will continue to grow and to exert increasing influence. Significantly, most members of this group are also committed to the principle that Antarctica forms part of the common heritage of mankind; and have recognized the need to cooperate on Antarctic affairs (especially environmental issues) with other international bodies and associations (such as Greenpeace).

In a few short years, then, the current Treaty may terminate. Which of the above contending groups will win is anybody's guess—although the tide of history is perhaps in favour of the common heritage concept and the countries that support this principle. One thing, however, is crystal-clear. A sound and stable dispensation must be found to replace the *modus operandi* of the current system: for without such political stability, the inherent risk factors for a mining venture in such a simmering environment would be too great to contemplate as a viable proposition.

The second major issue is the environmental question. It is clear that Antarctica has vast resources and—also—that those resources *can* be tapped economically (for the experience of mining in Arctic zones has demonstrated this beyond all serious doubt). But the environmental issue remains. Antarctic exploitation promises tempting short-term profits; but the long-term consequences may outweigh any such gains. The whaling industry alone provides a salutary warning: and interference in the delicate balance of Antarctica, given the continent's vital role in world weather and the global food-chain, may well prove ecologically disastrous. The environmentalists are therefore fixed in their resolve. Antarctica must not be mined—and especially not for short-term profit. There remains, however, yet another factor to take into account in weighing the environmental issue. Pollution is a major problem in the world today and is likely to increase as industrialization gains pace throughout the world. Platinum offers one important means of controlling such pollution; and it may be that the largest deposits of this vital metal are split between Antarctica and South Africa. This geological speculation raises important additional considerations.

The first is simply stated. Environmental concern to protect against pollution, using platinum, may outweigh the fear of exploiting Antarctica and the consequent hazards to the world's ecology. This may indicate, accordingly, that the mining of platinum (if not of other minerals) should indeed be pursued in Antarctica.

The second factor is more complex. It is well known (and has been broadly documented in this study) that the South African Government's conduct in relation to the majority of the country's inhabitants falls far below accepted international standards for the protection of human rights. It is possible that pressure could be brought to bear upon Pretoria to change its racial policies by a threat to stop buying platinum from South Africa and obtain supplies from Antarctica instead.

To summarize, then, there are important factors both for and against the mining of Antarctica. Against such exploitation are:

- the cost and difficulty of mining in its harsh conditions;
- the political uncertainty as to what will happen on the termination of the first 30-year phase of the Treaty; and
- the environmentalist concern to protect Antarctica (and the world) from further interference in the global ecosystem.

Against these considerations must, however, be weighted the factors in favour of, or conducive to, Antarctic exploitation. These are:

- the wealth of the continent's resources for the benefit of all mankind;
- the growing international consensus that Antarctica forms part of the common heritage of mankind and should be placed under United Nations' jurisdiction;
- the potential to control global industrial pollution through Antarctic platinum; and
- the possibility of compelling positive change in 'anti-social' countries, such as South Africa, through reliance on Antarctic minerals instead of those available from countries of this kind.

What course should be pursued? It is not the purpose of this book to prescribe appropriate action, but rather to explore the possibilities and open up the topic for discussion. Yet—if the writer may express a personal view—exploitation should not lightly be undertaken: not at least without far clearer understanding of the environmental costs and consequences. At the same time, however, Antarctica's potential (especially to supply the world with platinum) should be acknowledged and, if need be, tapped in order to control pollution and to prevent 'anti-social' conduct on the part of nations—both of which may pose even greater threats to the well-being of the world and all its peoples. In theory these goals could in fact be attained without ever proceeding with active mining in Antarctica, provided (1) a firm knowledge of Antarctica's resource inventory were available, and (2) the costs of exploiting these resources were accurately known. With such a scientific data base the global community could exploit Antarctica's natural resources passively and force global mineral owners to mine their resources more diligently, more economically, and with better socio-political standards.

This may seem Utopian: but with resolution, hard research, and real commitment to greater communication and understanding between all countries, it may well be within our grasp. If this deal is unattainable, however, then—at minimum—the present Antarctic Treaty nations must be willing to surrender their claims to sovereignty and overcome their distrust of other states: and thus enable the recognition of Antarctica as part of the 'common heritage' of mankind, to be used for the true benefit of all inhabitants of this dynamic globe.

APPENDICES

A1. THE ANTARCTIC TREATY—A SUMMARY

(Modified from National Science Foundation USA sources)

A rigorous analysis of the Antarctic Treaty can be found in Auburn (1982) and in Australian Department of Foreign Affairs (1983), whilst Bush (1982) provides the most complete collection of International and National documents which are openly available. Mitchell and Tinker (1980), Mitchell (1983) and Kimball (1983b), provide adequate background for a more general executive overview. *The handbook of measures in furtherance of the principles and objectives of the Antarctic Treaty* (1983, 3rd edition, 222 pp.), contains all the recommendations adopted by Treaty parties, copies of the Antarctic Treaty, the Convention on the Conservation of Antarctic Marine Living Resources, Treaty rules of procedures and other basic documents. (This is a public document: contact Treaty Party Governments.) The name of this document is to be changed to *Handbook of the Antarctic Treaty,* and it is anticipated that the history of the Treaty and other information will be included. It will be updated following each consultative meeting.

Preamble

The Governments of Argentina, Australia, Belgium, Chile, the French Republic, Japan, New Zealand, Norway, the Union of South Africa, the Union of Soviet Socialist Republics, the United Kingdom of Great Britain and Northern Ireland, and the United States of America;

Recognize that it is in the interest of all mankind that Antarctica shall continue forever to be used exclusively for peaceful purposes and shall not become the scene or object of international discord.

Acknowledge the substantial contributions to scientific knowledge resulting from international cooperation in scientific investigation in Antarctica.

Are convinced that the establishment of a firm foundation for the continuation and development of such cooperation on the basis of freedom of scientific investigation in Antarctica as applied during the International Geophysical Year accords with the interests of science and the progress of all mankind.

Are convinced also that a treaty ensuring the use of Antarctica for peaceful purposes only and the continuance of international harmony in Antarctica will further the purposes and principles embodied in the Charter of the United Nations.

Antarctic Treaty provisions—14 articles

1. Apply everywhere South of 60°S, but do not prejudice exercise of high seas rights under International Law.
2. No military installations or activities (except in support of science).
3. Right of inspection of scientific basis and full exchange of information.
4. Freedom of scientific access and scientific co-operation.
5. Agree to disagree on claims and freeze positions thereon.
6. Prohibit nuclear explosions or nuclear waste disposal.
7. Peaceful resolution of any jurisdictional disputes.
8. Consult regularly on measures to further Treaty principles.
9. Modify or amend only by unanimity of Consultative parties in first 30 years (to 23/6/91); thereafter by majority of Consultative Parties.
10. Unanimously approve recommendations of Consultative Meetings and when ratified, augment and expand the Treaty provisions.

The Antarctic Treaty is open for accession by any state which is a member of the United Nations and to others upon invitation with the consent of all the Antarctic Treaty Consultative parties (ATCP).

Treaty consultative meeting recommendations (up to 1983)

One hundred and thirty-one thus far issued on measures related to:

(a) uses of the Antarctic (24 measures);
(b) preservation and conservation of wildlife and living resources (37);
(c) facilitation of scientific research (10);
(d) facilitation of international scientific co-operation (21);
(e) exchanges of information (15);
(f) operation of the Treaty and Consultative meetings (22);
(g) postal services (2).

By-products of the Treaty (e.g. part of the Treaty System)

1. *Convention for the Conservation of Antarctic Seals.* Negotiated among Treaty Parties but open for signature by non-treaty parties as a separate and independent agreement, opened for signature in February 1972; entered into force in March 1978. Ratified by 10 Antarctic Treaty Consultative parties.

2. *Convention on the Conservation of Antarctic Marine Living Resources* (C.C.A.M.L.R.). Opened for signature in 1980. Ratified by Argentina, Australia, Chile, FRG, GDR, Japan, New Zealand, South Africa, UK, USA, and USSR and entered into force 7 April 1982. First meeting of Commission held in Hobart, Australia, 24 May–11 June 1982 with all above (less Argentina, plus EEC) as members (Argentina's membership effective 28 June 1982; the EEC acceded on 2 April 1982) and the following countries as observers: Argentina, Belgium, France, Norway, and Poland. Secretariat established at Hobart, Australia. Scientific Committee failed to adopt rules of procedure. Budgets adopted for FY82 ($472.8K) and FY83 ($604.45K). Executive Secretary appointed: Dr Darry Powell of Australia.

3. *Regime Concerning Exploration and Exploitation of Antarctic Mineral Resources.* Initial session of Special Consultative Meetings (SCM) held at Wellington, NZ in June 1982. A second session was held at Bonn, FRG, June 1983, followed by a third in Washington USA, January 1984, a fourth in Tokyo, May 1984, and a fifth meeting in Rio de Janeiro, February 1985.

Parties to the Antarctic Treaty

1. Twelve Original Signatories—Argentina, Australia, Belgium, Chile, France, Japan, New Zealand, Norway, South Africa, USSR, UK, and USA.

2. Sixteen Acceding States—Czechoslovakia, Denmark, The Netherlands, Romania, German Democratic Republic, Bulgaria, Uruguay, Italy (3.18.81), Peru (4.10.81), Papua New Guinea (3.16.81), Spain (3.31.82), Peoples Republic of China (June 1983), Hungary, Sweden, Finland, Republic of Cuba (1984). These states have been granted observer status at Biennial Antarctic Treaty meetings since 1983.

3. Sixteen Consultative parties (voting members)—The 12 original signatories and three scientifically qualified accedents (Poland, FRG, and India), plus Brazil.

4. Thirty-two contracting parties—all of the above.

Operation of the Treaty

1. Periodic meetings open only to the Consultative Parties to develop recommendations. Twelve held so far: 1961 Canberra; 1962 Buenos Aires; 1964 Brussels; 1966 Santiago; 1968 Paris; 1970 Tokyo; 1972 Wellington; 1975 Oslo; 1977 London; 1979 Washington; 1981 Buenos Aires; 1983 Canberra. The thirteenth meeting will be held in Brussels in 1985.

2. Since 1983, observers from nations that have acceded to the Treaty have been invited to attend these meetings.

3. All recommendations require unanimous approval of all voting representatives at the consultative meeting.

4. Recommendations enter into force upon ratification by all of the consultative parties, and are applicable to all of Antarctica with no distinction between claimed and unclaimed regions.

5. US is the Depository Government—there is no Secretariat.

6. Special Consultative meetings for single issue resolution. Four have been held so far: (1) London 1977 to admit Poland to Consultative party Status; (2) Canberra/Buenos Aires 1978 then again Canberra 1980 to develop a Convention on the Conservation of Antarctic Marine Living Resources (signed in September 1980); (3) Buenos Aires 1981 to admit FRG to Consultative Status; (4) A Special Consultative Committee meets periodically for the purpose of negotiating a mineral resources regime. Initial sessions were held June 1982 at Wellington; June 1983 in Bonn; January 1984 in Washington; May 1984 in Tokyo; February 1985 in Rio de Janeiro.

7. Meetings of Group of Experts for specific problem analysis, such as logistics, telecommunications, environmental impacts, legal/political concerns, etc.

A2. NATIONAL INTERESTS IN ANTARCTICA
(Compiled from several sources)

Thirty-two nations are known to have expressed significant interest in Antarctica in recent years. These expressions of interest range from accession to the Antarctic Treaty (Sweden, Finland, and Cuba are the most recent additions) to the mounting of continuous, large expeditions to Antarctica and full participation in the Antarctic Treaty consultative process (the USSR and the USA are examples).

Of the 32 nations, 14 currently operate stations and/or research programmes on the continent of Antarctica. Since 1980, 12 of these 14 nations have commenced, or completed, substantial new investments in their Antarctic programmes. These investments range from the building of new ships to the complete refurbishment of existing stations or the establishment of year-round stations. These 12 nations are Argentina, Australia, Federative Republic of Brazil, Chile, the Federal Republic of Germany, France, India, Japan, New Zealand, the Union of Soviet Socialist Republics, the United Kingdom, and the United States of America.

The other two nations with current research expeditions to Antarctica have continued their programmes at levels comparable to former years. These nations are the Republic of South Africa, and Poland.

During the 1982–1983 Antarctic season, 35 permanent scientific stations were operated by 891 winter-over personnel. Additionally 20 summer stations were manned, and the augmented summer personnel numbered about 2050. Thirty-six ships supported the scientific and logistic activities.

Argentina

Argentina maintains eight year-round stations, all of them coastal: six along the Antarctic Peninsula and two on the Weddell Sea Coast. In April 1984, one of the peninsula stations, Almirante Brown, burned to the ground. The station residents were evacuated by US research ship *Hero*. Wintering personnel totalled 227 in 1981. Argentina also operates six summer stations: summer-augmented personnel number about 160. The army and navy provide almost all logistics and support. Aeroplanes and helicopters are operated, and one station, Marambio, has a runway that C-130s use year-round. Science is performed in most of the Antarctic disciplines, marine science is carried out in Antarctic waters and the Drake Passage. The southernmost Argentinian town Ushuaia is sometimes used by USARP as a stepping stone to support their operations in the Antarctic Peninsula.

Since 1981 Argentina has purchased two new ships for Antarctic service. *Bahia Paraiso* is a supply ship and *Almirante Irizar* a large $45-million icebreaker, equipped for marine research. *Peurto Deseado* is a new research ship that will see limited Antarctic service. Argentina has maintained a field programme and a manned station in Antarctica continuously since 1904. In 1937, Argentina extended claims to all the territories of the Falkland Islands Dependencies, which included part of the Antarctic mainland. In 1943 the country asserted a territorial claim to the Antarctic Peninsula and adjacent areas by means of Decree Law No. 2191, which re-establishes the 'National Territory of Tierra del Fuego, the Antarctic, and the Islands of the South Atlantic'.

In 1978 the wife of an Argentinian officer gave birth to the first baby ever born in Antarctica. Station officials routinely ask visitors to present passports. Argentina is an Original Consultative Party to the Antarctic Treaty.

Argentina is hopeful that the headquarters of the Mineral Resource Regime, now under negotiation, will be located in Mar del Plata. Argentina co-operates actively in Antarctica with other nations.

Australia

Australia operates three year-round stations on the coast of East Antarctica with about 90 wintering personnel. Research projects are well distributed among the disciplines, and the country has a reputation for excellent work in glaciology and meteorology. In summer an aeroplane and helicopters support field parties that range inland. Intercontinental transport and resupply are by three chartered ice-strengthened ships.

Following a visit to Casey Station in 1980–91 by government representatives, a $58 million 10-year redevelopment programme for Australia's Antarctic bases was approved. There will be a shift in emphasis, increase in size, and improvement of the quality of Australia's research programme. A major new programme in southern ocean marine biology will be initiated.

Headquarters for the Antarctic programme were shifted in 1980 from Melbourne to a new building in Hobart, a move that received national attention. Hobart also has been selected as sites for the Commission and Secretariat Headquarters of the Convention on the Conservation of Antarctic Marine Living Resources.

Australia's territorial claim is the largest in Antarctica. Australia is an Original Consultative Party to the Antarctic Treaty.

Belgium

Belgium is signatory to the Antarctic Treaty and continues to participate in Antarctic Treaty Consultative Meetings. The country was active in Antarctica during the International Geophysical Year and maintained a station on the coast of Queen Maud Land until 1951. It has not conducted a field programme since joint Belgian and South African geological–geophysical expeditions in the three seasons from 1967 to 1970.

Federative Republic of Brazil

In the mid-1970s, Brazil began planning an expedition to Antarctica, but the expedition did not take place. A resurgence of activity took place in 1981 when Brazil purchased the supply ship *Thala Dan* from Denmark. Seven foreign scientists from five countries attended a seminar in São Paulo in the autumn of 1982 to discuss the designing of a Brazilian Antarctic programme. The Brazilians carried out oceanographic research in 1982–1983, mainly in the Bransfield Strait area:13 scientists were aboard the research vessel of São Paulo University. In 1983–1984, the Brazilians continued with their oceanographic work and established a base, Comandante Ferray, on King George Island which is situated on the west side of the Antarctic Peninsula. Brazil was granted consultative party status in September 1983. Brazil participates in the BIOMASS program. The country applied for membership in SCAR in July 1984.

Bulgaria

Bulgaria acceded to the Antarctic Treaty in 1978.

Chile

Chile has three year-round stations along the Antarctic peninsula and operates four additional stations in the summer. Wintering personnel total about 50; summer personnel about 230. Three ships aided by helicopters support a research programme that centres on marine biology. A new airfield, called Teniente Marsh Base, was built near the meteorological station, Presidente Frei, on King George Island for intercontinental operations.

Two Air Force Officers visited McMurdo in December 1981 to study C-130 air drop techniques. Chilean aircraft have commenced regular operations from Punta Arenas to Teniente Marsh Base on King George Island. The airport and harbour facilities of the southern Chilean town Punta Arenas are sometimes used by USARP as a stepping stone to support their operations in the Antarctic Peninsula, and more recently in the Thiel Mountains. In October 1981, a US scientific team was transported from Chile to Marsh by a Chilean C-130 and onward to a field site by Chilean helicopter. This co-operative effort continued in 1982–83 and 1983–84 and reduced the US programme's need of scheduling special trips by the R/V *Hero.* Three Chileans visited the US programme in 1982–83. In October 1982 Chile hosted a multinational seminar on Antarctic resource policy at its Teniente Marsh Base.

Chile asserted a territorial claim to the Antarctic Peninsula lying between 53° and 90°W and adjacent areas in 1940 by Presidential Decree (Territorio Antartico Chileno). This very largely overlapped the areas already claimed by both Britain and Argentina. The nation is an Original Consultative Party to the Antarctic Treaty.

Czechoslovakia

Czechoslovakia acceded to the Antarctic Treaty in 1962.

Denmark

Denmark acceded to the Antarctic Treaty in 1965.

Federal Republic of Germany

The Federal Republic of Germany resumed German expeditions to Antarctica during the 1975–1976 and 1977–1978 summer seasons with ship-based biological and geophysical research.

In the 1979–80 season the chartered motor ship *Polarsirkel* supported a site survey on the Filchner and Ronne Ice Shelves for a new, year-round station and supported investigations in glaciology, biology, and oceanography.

During 1979–80, in GABOVEX I the motor ship *Schepelsturm* supported geophysical, petrological and geological studies in the Ross Sea and in northern Victoria Land. Prefabricated shelters were erected for future use at 71°S 154°E. During GANOVEX II in 1981–1982 the support vessel *Gotland II* was crushed by ice and lost. GANOVEX III took place in 1982–1983 using another chartered vessel for support. GANOVEX IV is planned for 1984–1985.

The country has built a multipurpose Antarctic ice-strengthened ship, the *Polar Stern*, 360 ft long, that has two helicopters, a crew of 35, and space for 40 scientists and 26 additional passengers. It cost $105.5 million, less helicopters. Two Dornier aircraft have recently been acquired and will be tested in 1984–1985 for, *inter alia*, airborne geophysical studies over Dronning Maud Land. Extensive geophysical investigations are also planned for in the Weddell Sea region in 1985–1986.

A wintering-over station, Georg von Neumayer, was established in 1980–81 at 70°37′S 08°22′W. The FRG operates three summer stations with about 60 personnel. Winter-over personnel number nine.

The government earmarked some DM 380 million for its Antarctic programme for the period 1982–1983.

The Federal Republic of Germany acceded to the Antarctic Treaty in 1979 and gained consultative status in March 1981.

Finland

Finland acceded to the Antarctic Treaty in 1984.

France

France has one station in Antarctica, operated year-round by a wintering party of about 30 on the coast of Adelie Land. Resupply of the station is carried out by chartered vessel. The scientific programme emphasizes glaciology, with traverses and deep ice drilling, but includes all disciplines. In recent years France has collaborated with the United States in glaciology and in the installation of automatic weather stations. In concert with efforts of other European nations, such as the UK, FRG, and Norway, French scientists propose to start an integrated marine geophysical survey in the Weddell Sea in 1986.

France is concentrating on the building of a wheeled runway at its year-round station, Dumont d-Urville (66°40′S, 140°01′E). Serious environmental concerns have been expressed internationally about the effects on the local fauna.

By Presidential Decree in 1924, France claimed the sector of Antarctica in which its station lies. The nation is an Original Consultative Party to the Antarctic Treaty.

German Democratic Republic

East Germany acceded to the Antarctic Treaty in 1974. East Germans participate frequently in Soviet Expeditions and have established summer stations for independent field programmes. GDR is a member of CCAMLR Commission and of SCAR.

Hungary

Acceded to the Treaty in 1984.

India

In January 1982 a 20-member Indian team supported by the chartered vessel *Polarsirkel* and helicopters landed in Antarctica and set up a temporary camp ‘Southern Ganges’ at 70°03′S 41°02′E, on land claimed by Norway. India has since established a permanent camp at 70°57′S, 12°49′E. India acceded to the Antarctic treaty in August 1983. India mounted a large second expedition (28 scientists) in the 1982–1983 season, and a third expedition in the 1983–1984 season. In August 1983, India was granted consultative party status. It has an Antarctic budget of about $2 million. India has welcomed the UN objective study on all aspects of Antarctica, as called for during a 1983 UN General Assembly debate.

Italy

Italy acceded to the Treaty in the spring of 1981. Italy has asked for information on organizational, technical, administrative and logistical guidelines to conduct Antarctic exploration, and the US has provided the materials. Two Italians visited the New Zealand programme in 1982–1983. Italy was an observer at SCAR in 1982.

Japan

Japan's stations, Syowa and Mizhuho, in East Antarctica are operated year-round by about 35 people. Extra summer personnel reach about 20. Station supply is by an icebreaker operated by the Maritime Self Defence Force. The new icebreaker of 12 000 tons, the *Shirase*, replacing the aging *Fuji*, was completed in November 1982, and sailed south in 1983–1984.

Meteorology and ionospheric physics are emphasized, but Japan also does extensive research on marine biology. The objective is to get scientific knowledge as a basis for assessment of the resource potential.

Japan's Natural Resources and Energy Agency has undertaken a multi-year, ship-based assessment of the potential for mineral resources on the Antarctic continental shelf; areas of emphasis are the Bellingshausen Sea (1980–81), the Weddell Sea (1981–82), and the Ross Sea (1982–83). Current plans call for a three-year extension of the survey around the coast. The assessment includes bottom sampling, depth sounding, seismic reflections, sonar buoy seismic refraction, magnetometry, gravimetry, and terrestrial heat flow measurements.

Japan continues the upgrading of her facilities which now include data acquisition via satellite links.

Japan and the United States have collaborated extensively in recent years in geological and geophysical research and in the collection of meteorites, mainly in the McMurdo Sound region. Three US scientists from the University of Alaska carried out research aboard *Fuji* in 1982–83.

Japan is an Original Consultative Party to the Antarctic Treaty.

Netherlands

The Netherlands acceded to the Antarctic treaty in 1967. The Netherlands has had scientists accompany Belgian expeditions. The Netherlands was an observer at SCAR in 1982. The Netherlands is negotiating an Antarctic expedition in the near future, using FRG logistics support.

New Zealand

New Zealand maintains one year-round station, Scott Base (77°51′S, 166°46′E), located on Ross Island just 3 km from the US McMurdo Station, and one summer station, Vanda, in the dry valleys of southern Victoria Land. New Zealand provides for several C-130 aeroplane round trips between Christchurch and McMurdo, aircraft crews at McMurdo and other services each year, in return for transporation to and within Antarctica provided by the United States during the summer season. As a result, New Zealand is able to augment its wintering party of about 11 with a large number of summer personnel. 190 science and support personnel were involved in the 1982–83 programme. New Zealand and US parties frequently perform co-operative research projects in the austral summer. The New Zealand field programme includes glaciology, biology, geophysics, geology, and topographical surveys and mapping.

New Zealand has operated continuously in Antarctica since the International Geophysical Year and is nearing the completion of a rebuilding programme of Scott Base. New Zealand operates on a budget of about US$5 million (1983).

New Zealand is embarking on an eight year drilling programme west of McMurdo Sound to obtain cores for dating. Interest continues in obtaining a ship for supply and research and in developing future programmes in marine biology, possibly using Hallett Station (near Cape Adare, northern Victoria Land).

New Zealand is an Original Consultative Party to the Antarctic Treaty. The New Zealand territorial claim was asserted in 1923 by a British Order in Council that appointed the Governor General of New Zealand as Governor of the Ross Dependency, the name applied by New Zealand to its Antarctic claim.

Norway

Norway operated one station in Antarctica, on the coast of Queen Maud Land, during the International Geophysical Year and continued its research through 1959. In 1974–75 the United States provided air support to a Norwegian geological party. During the 1976–77 season, a Norwegian ship, the *Polarsirkel,* operated in Antarctic waters and supported two summer stations. The recent scientific work has focused on oceanography, geology, and geophysics. Norway maintained no research stations in the 1982–1983 Antarctic season. An integrated marine geophysical programme has been proposed for the 1983–84 season in the southern Weddell Sea.

Norway's territorial claim was made by Royal Proclamation in 1939. The nation is an Original Consultative Party to the Antarctic Treaty and continues to participate in Antarctic Treaty Consultative Meetings.

Papua New Guinea

Papua New Guinea succeeded to the Antarctic Treaty in the spring of 1981: when the country gained independence from Australia, it could have seceded from the Treaty, but it opted to continue membership.

People's Republic of China

Two Chinese observers visited US, New Zealand, and Australian stations in Antarctica during the 1979–80 season. The Chinese also visited the 1981–82 programmes of New Zealand and Australia, and PRC scientists participated in New Zealand's 1982–83 programme, and wintered over at Australian stations.

China has indicated an interest in mounting an Antarctic expedition and is considering the establishment of a permanent base on the continent. A delegation from the PRC visited Washington D.C. in July 1984, and PRC scientists visiting US academic institutions have participated in Antarctic field research as members of their US host scientists' projects. PRC is planning a 48-person ship expedition to the Antarctic Peninsula in the 1984–1985 season. PRC is quite in favour of United Nations involvement in Antarctica.

China acceded to the Treaty in 1983.

Peru

Peru acceded to the Antarctic treaty in the spring of 1981. Two Peruvian officers, one Coast Guard and one Navy officer accompanied the USCG *Polar Sea* on her 1981–92 Antarctic deployment. Peru expressed full support for the 1983 UN resolution for a United Nations objective study on all aspects of Antarctica.

Poland

Poland acceded to the Antarctic Treaty in 1961 and briefly operated a coastal station in East Antarctica with Soviet support.

In 1977 Poland established a year-round station. Henrik Arctowski (62°09′S, 58°28′W), on King George Island near the northern tip of the Antarctic Peninsula, using the research ship, *Professor Siedlecki,* and the factory trawler ship *Tazar* for support and for biological and fisheries research. At the station, research is performed in most of the Antarctic scientific disciplines. About nine personnel winter over. In 1975, Poland accepted title to the former Soviet summer station Oasis, renaming it Dobrowolski.

Poland attained Antarctic Treaty consultative status in 1977 and has participated in Consultative meetings since that time.

Republic of Cuba

Cuba acceded to the Treaty in 1984.

Republic of Korea

Fishing activities were carried out in Antarctic waters in 1978.

Republic of South Africa

South Africa's station Sanae (70°18′S, 02°26′W, previously known as Norway station), on the Fimbul Ice Shelf, along the coast of Queen Maud Land, has operated since the International Geophysical Year; the current wintering crew numbers about 16 persons. Research at Sanae is mainly in the atmospheric sciences. It also supports active earth and biological science summer programmes. A new summer station in the vicinity of Grunehogna (71°02′S, 02°48′W) was erected during the 1982–83 season. About 10 scientists occupy this site and are supported by two helicopters with a range of about 300 km. In 1978, the country placed into service an ice-strengthened resupply and research vessel, the MV SA *Agulhas* for the resupply of Sanae and the sub-Antarctic islands. This vessel has space to accommodate scientists from other nations in co-operative programmes. A second, newer vessel, the *Africana,* is a 180 ft long ice-strengthened fisheries research ship designed primarily for biological oceanography. No expansion of present involvement and funding is planned in the immediate future. Plans are under way for co-operative geophysical surveys in the Weddell Sea–Queen Maud Land offshore environment between South Africa and the Federal Republic of Germany.

South Africa is an Original Consultative Party to the Antarctic Treaty.

Romania

Romania acceded to the Antarctic Treaty in 1971.

Saudi Arabia

In the 1970s and 1980 Saudi Arabia provided financial support for studies and conferences regarding the feasibility of towing Antarctic icebergs for use as fresh water in countries in the temperate latitudes.

Spain

Spain acceded to the Antarctic Treaty 31 March 1982.

Sweden

Sweden acceded to the Antarctic Treaty in 1984.

Taiwan

Taiwan carried out experimental fishing in Antarctic waters in 1976–1978. In 1977, a 700-ton vessel fished in the waters off Enderby Land for 18 days. The catch was approximately 130 tons.

Union of Soviet Socialist Republics

The Soviet effort in Antarctica is substantial and approximates that of the United States. The Soviet Union has one station in the interior—Vostok (78°28′S, 106°48′W), at the geomagnetic south pole. In addition, the Soviets have six active year-round coastal stations located in different sextants of the continent: Mirnyy (66°33′S, 93°01′E, established in 1956); Novolazarevskaya (70°46′S, 11°50′E, established in 1961); Molodezhnaya (67°40′S, 45°51′E, established in 1962); Bellingshausen (62°12′S, 58°56′W, established in 1968); Leningradskaya (69°30′S, 159°23′E, established in 1970); and Russkaya (74°46′S, 136°52′W, established in 1979). The Soviet wintering parties total some 300 persons, more than twice the US number. Because transportation of Soviet personnel (and supplies) to Antarctica is almost entirely by ship, the summer research contingent is essentially the same as the wintering-over. The USSR operates three summer stations; in the 1982–1983 season, extra summer personnel was about 127.

Annual resupply generally is performed by six to ten ships, of which two are research vessels carrying a large scientific complement. The Soviets successfully completed one intercontinental air transport of personnel in 1979 from Madagascar, landing on a snow-compacted runway at Molodezhnaya. Air-transportation is also conducted from Maputo in Mozambique. Within Antarctica, air transport is provided by airplanes equivalent to DC-3s and by large helicopters. Two IL-18D aircraft landings were made at Molodezhnaya in 1981–82.

Soviet Antarctic research is concentrated in upper atmospheric physics, meteorology, and earth sciences. The terrestrial life sciences receive little emphasis, but marine biology and fisheries research are emphasized. In the 1981–82 season, the Soviet ship *Somov* supported a US research team on the Weddell Polynya Expedition.

The Soviets operate an active fishing fleet in and near Antarctic waters. At a recent Hobart meeting, the Soviets stated that their previous year's krill catch was 500 000 metric tons.

Like the United States, the Soviet Union has performed research in Antarctica continuously since the International Geophysical Year. Also like the United States, the Soviet Union makes no Antarctic territorial claim and recognizes none of the claims made by other countries. The Soviet Union is an Original Consultative party to the Antarctic Treaty.

A Soviet exchange scientist wintered at South Pole Station in 1982.

United Kingdom

The British Antarctic Survey maintains four year-round stations, one on the Weddell Sea coast, one in the South Orkney Islands and two along the Antarctic Peninsula. Wintering parties totalled 62 in the 1984 winter. Annual resupply uses two ships plus an ice patrol ship. Twin otter airplanes support summer field operations. Science is performed in all the Antarctic disciplines. Geology, geophysics, glaciology, and biology during the 1982–83 summer involved an additional 42 people. The UK inaugurated a continuing Antarctic Research programme in 1925. Today it operates on an annual budget of about US$ 15 million.

Three ships regularly operate in Antarctic waters: RRS *Bransfield* is principally involved with the resupply of the Antarctic stations; RRS *John Biscoe* has been extensively refitted to conduct marine biological and oceanographic research; HMS *Endurance*, the Royal Navy's ice patrol vessel, assists with some scientific programmes, particularly where helicopters are needed. Marine geophysical research is carried out from a chartered vessel and may be supported by the aforementioned vessels.

The British Antarctic Survey recently received a substantial increase in funds, largely as a result of the situation with the Falkland Islands in early 1982, to expand its scientific programmes. This has led to enlarging and upgrading its Antarctic stations, including the rebuilding of Halley station using revolutionary plywood tubes to house the accommodation buildings.

British Antarctic Territory, the UK claim in Antarctica since 1908, includes the Antarctic Peninsula and adjacent areas. The United Kingdom is an Original Consultative Party to the Antarctic Treaty.

United States of America

The United States maintains four year-round stations in Antarctica: McMurdo, the logistics colony on Ross Island; Amundsen–Scott South Pole, at 90°S latitude; Palmer, on Anvers Island off the western coast of the Antarctic Peninsula, and Siple Station in Ellsworth Land (which was closed for the 1984 winter). Wintering personnel in 1982–1983 numbered 123; this was augmented by 1126 US summer personnel. Forty foreign nationals participated in the US summer programme.

Resupply of US stations is accomplished through McMurdo by a cargo ship and a tanker, preceded by icebreakers. Cargo and personnel are moved from McMurdo to the inland stations by air. Palmer is resupplied by icebreaker and the 219 ft chartered ice-strengthened ship *Polar Duke*. This ship is well equipped with research laboratories and winches for oceanographic work. Airplanes operate October through February between New Zealand and McMurdo.

A capability unique to the US Antarctic Research Program (USARP) is the use of large ski-equipped aeroplanes to place summer research camps at almost any location in Antarctica. Helicopters can be transported inside the aeroplanes to these sites and used for increased local mobility.

Research is supported in a balanced programme comprising biological and medical research, the ocean sciences, upper atmosphere physics, meteorology, glaciology, and the earth sciences.

The United States reserves the right to, but does not make a territorial claim in Antarctica, and recognizes none of the claims by other nations. The United States is an Original Consultative party to the Antarctic Treaty, and is the Treaty's depository government.

Uruguay

Uruguay acceded to the Antarctic Treaty in 1980. Two Uruguayan air force officers visited the US programme in Antarctica in 1981–82.

A3. ANTARCTIC MINERAL RESOURCES

Resolutions of the Antarctic Consultative Meeting XI

The Representatives

Recalling the provisions of the Antarctic Treaty, which established a regime for international cooperation in Antarctica, with the objective of ensuring that Antarctica should continue forever to be used exclusively for peaceful purposes and should not become the scene or object of international discord;

Convinced that the framework established by the Antarctic Treaty has proved effective in promoting international harmony in furtherance of the purposes and principles of the United Nations Charter, in prohibiting *inter alia* any measures of a military nature, in ensuring the protection of the Antarctic environment, in preventing any nuclear explosions and the disposal of any radioactive waste material in Antarctica, and in promoting freedom of scientific research in Antarctica, to the benefit of all mankind;

Convinced further of the necessity of maintaining the Antarctic Treaty in its entirety and believing that the early conclusion of a regime for Antarctic mineral resources would further strengthen the Antarctic Treaty framework;

Desiring without prejudice to Article IV of the Antarctic Treaty to negotiate with the full participation of all the Consultative parties to the Antarctic Treaty an appropriate set of rules for the exploration and exploitation of Antarctic mineral resources;

Noting the unity between the continent of Antarctica and its adjacent offshore areas;

Mindful of the negotiations that are taking place in the Third United Nations Conference on the Law of the Sea;

Reaffirming their commitment to the early conclusion of a regime for Antarctic mineral resources which would take due account of the respective interests of the Consultative Parties as regards the form and content of the regime, including decision-making procedures, as well as the special characteristics of the Antarctic area;

Recalling Recommendations VII-6, VIII-14, IX-1 and X-1;
Recalling further Recommendations VI-4, VII-1, VIII-11, VIII-13, IX-5, IX-6 and X-7.

Recommend to their Governments that:

1. They take note of the progress made toward the timely adoption of a regime for Antarctic mineral resources at the Eleventh Consultative meeting and related meetings and the importance of this progress.

2. A regime on Antarctic mineral resources should be concluded as a matter of urgency.

3. A Special Consultative meeting should be convened in order:

(a) to elaborate a regime;

(b) to determine the form of the regime including the question as to whether an international instrument such as a convention is necessary;

(c) to establish a schedule for negotiations, using informal meetings and sessions of the Special consultative Meeting as appropriate; and

(d) to take any other steps that may be necessary to facilitate the conclusion of the regime, including a decision as to the procedure for its adoption.

4. The Special Consultative meeting should base its work on this Recommendation and relevant Recommendations and reports of the Eighth, Ninth and Tenth Antarctic Treaty Consultative meetings.

5. The regime should be based on the following principles:

(a) the Consultative Parties should continue to play an active and responsible role in dealing with the question of Antarctic mineral resources;

(b) the Antarctic Treaty must be maintained in its entirety;

(c) protection of the unique Antarctic environment and of its dependent ecosystems should be a basic consideration;

(d) the Consultative Parties, in dealing with the question of mineral resources in Antarctica, should not prejudice the interests of all mankind in Antarctica;

(e) the provisions of Article IV of the Antarctic Treaty should not be affected by the regime. It should ensure that the principles embodied in Article IV are safeguarded in application to the area covered by the Antarctic Treaty.

6. Any agreement that may be reached on a regime for mineral exploration and exploitation in Antarctica elaborated by the Consultative Parties should be acceptable and be without prejudice to those States which have previously asserted rights of or claims to territorial sovereignty in Antarctica as well as to those States which neither recognize such rights of or claims to territorial sovereignty in Antarctica nor, under the provisions of the Antarctic Treaty, assert such rights or claims.

7. The regime should *inter alia*:

(a) Include means for:

(i) assessing the possible impact of mineral resource activity on the Antarctic environment in order to provide for informed decision-making;

(ii) determining whether mineral resource activities will be acceptable;

(iv) governing the ecological, technological, political, legal and economic aspects of those activities in cases where they would be determined acceptable, including:

the establishment, as an important part of the regime, of rules relating to the protection of the Antarctic environment: and

the requirement that mineral resource activities undertaken pursuant to the regime be undertaken in compliance with such rules.

(b) Include procedures for adherence by states other than the Consultative Parties, either through the Antarctic Treaty or otherwise, which would:

(i) ensure that the adhering State is bound by the basic provisions of the Antarctic Treaty, in particular Articles I, IV, V and VI, and by the relevant Recommendations adopted by the Consultative Parties; and

(ii) make entities of that State eligible to participate in mineral resource activities under the regime.

(c) Include provisions for cooperative arrangements between the regime and other relevant international organisations.

(d) Apply to all mineral resource activities taking place on the Antarctic continent and its adjacent offshore areas but without encroachment on the deep seabed. The precise limits of the area of application would be determined in the elaboration of the regime.

(e) Include provisions to ensure that the special responsibilities of the Consultative parties in respect of the environment in the Antarctic Treaty area are protected, taking into account responsibilities which may be exercised in the area by other international organizations.

(f) Cover commercial exploration (activities related to minerals involving, in general, retention of proprietary data and/or non-scientific exploratory drilling) and exploitation (commercial development and production).

(g) Promote the conduct of research necessary to make environmental and resource management decisions which would be required.

8. They promote and co-operate in scientific investigations which facilitate the effective operation of the regime taking into account *inter alia,* the relevant parts of the Report of Ecological, Technological and other Related Experts on Mineral Exploration and Exploitation in Antarctica (Washington, June 1979), attached as an annex to the Report of the Tenth Consultative Meeting.

9. With a view to improving predictions of the environmental impacts of activities, events and technologies associated with mineral resource exploration and exploitation should such occur, they continue with the assistance of the Scientific Committee on Antarctic Research to define programs with the objectives of:

(a) Retrieving and analysing relevant information from past observations and research programmes;

(b) Ensuring in relation to the needs for information identified by the Experts report, that effective use is made of existing programmes;

(c) Identifying and developing new programmes that should have priority, taking account of the length of time required for results to become available.

10. In elaborating the regime, they take account of the provisions of Recommendation IX-1, paragraph 8.

A4. DUFEK PLATINUM METALS MINE CASH FLOW EXAMPLE

MECON (MINERAL EVALUATION SYSTEM VERSION 1.1 - 81/4/30) DATE 83/06/23, PAGE 2

ECONOMIC MODEL OF ANTARTICA PLATINUM METALS MINE

PD/PT=0.42 REAL PT PRICE RISE 2PC 10 PC INFLATN****NOTE .1000E+04 = 1000.0****

		YEAR 84	YEAR 85	YEAR 86	YEAR 87	YEAR 88	YEAR 89	YEAR 90	YEAR 91
*C*PT PRICE 1983 US D/TROY OZ		.4845E+03	.4942E+03	.5041E+03	.5142E+03	.5244E+03	.5349E+03	.5456E+03	.5565E+03
*C*PD PRICE 1983 US D/TROY OZ		.1326E+03	.1353E+03	.1380E+03	.1407E+03	.1435E+03	.1464E+03	.1493E+03	.1523E+03
*C*U/G PRODUCTION TONNES/YEAR		0	0	0	0	0	0	.5533E+06	.6079E+06
-C-U/G RESERVES TONNES (YR ST)		.1500E+08	.1500E+08	.1500E+08	.1500E+08	.1500E+08	.1500E+08	.1500E+08	.1455E+08
-C-U/G ROM GRADE GRAM PT/TONNE		0	0	0	0	0	0	.3941E+01	.3941E+01
TOTAL ANNUAL SALES REVENUE		0	0	0	0	0	0	.7759E+08	.9543E+08
***ANNUAL PERSONNEL COSTS	D	0	0	0	0	0	0	.2065E+08	.2346E+08
***ANNUAL SUPPLIES COSTS	D	0	0	0	0	0	0	.7670E+07	.9007E+07
***ANNUAL POWER COSTS	D	0	0	0	0	0	0	.3862E+07	.4538E+07
***ANNUAL TRANSPORT COSTS	D	0	0	0	0	0	0	.2111E+07	.2552E+07
***ANNUAL GEN + ADMIN COSTS	D	0	0	0	0	0	0	.1731E+08	.1904E+08
TOTAL ANNUAL OPERATING COSTS		0	0	0	0	0	0	.5160E+08	.5860E+08
***EXPLORATION CAPEX	D	.4686E+07	.5155E+07	0	0	0	0	0	0
***SHAFT AND EQUIPMENT CAPEX	D	0	0	.2390E+07	.1052E+08	.8677E+07	.6363E+07	0	0
***UNDERGROUND MINING CAPEX	D	0	0	.9608E+06	.1285E+08	.2944E+08	.4655E+07	0	0
***TREATMENT PLANT CAPEX	D	0	0	.2668E+07	.4179E+08	.5708E+08	.1403E+08	0	0
***GENERAL FACILITIES CAPEX	D	0	0	.1837E+07	.2878E+08	.3931E+08	.9665E+07	0	0
*** PROJECT OVERHEAD COSTS	D	.2812E+06	.3093E+06	.5248E+06	.6473E+07	.9212E+07	.2364E+07	0	0
***WORKING CAPITAL	D	0	0	0	0	0	0	.1720E+08	.2334E+07
TOTAL ANNUAL CAPITAL COSTS		.4967E+07	.5464E+07	.8381E+07	.1004E+09	.1437E+09	.3708E+08	.1720E+08	.2334E+07
ANNUAL RECOVERED CAPITAL COSTS		0	0	0	0	0	0	0	0
UNADJUSTED CAPITAL ALLOWANCES		0	0	0	0	0	0	.3000E+09	0
ANNUAL TAXATION ALLOWANCES		0	0	0	0	0	0	.2599E+08	.3683E+08
ANNUAL TAXATION AT WBK RATE		0	0	0	0	0	0	0	0
ANNUAL CASH FLOW		-.4967E+07	-.5464E+07	-.8381E+07	-.1004E+09	-.1437E+09	-.3708E+08	.8790E+07	.3449E+08
CUMMULATIVE CASH FLOW		-.4967E+07	-.1043E+08	-.1881E+08	-.1192E+09	-.2629E+09	-.3000E+09	-.2912E+09	-.2567E+09
PRESENT VALUES AT SELECTED DISCOUNT RATES									
DISCOUNT RATE 0 PERCENT		-.4516E+07	-.4516E+07	-.6297E+07	-.6859E+08	-.8923E+08	-.2093E+08	.4511E+07	.1609E+08
DISCOUNT RATE 5.0 PERCENT		-.4301E+07	-.4096E+07	-.5440E+07	-.5643E+08	-.6992E+08	-.1562E+08	.3206E+07	.1089E+08
DISCOUNT RATE 10.0 PERCENT		-.4105E+07	-.3732E+07	-.4731E+07	-.4685E+08	-.5541E+08	-.1182E+08	.2315E+07	.7507E+07
DISCOUNT RATE 15.0 PERCENT		-.3927E+07	-.3414E+07	-.4140E+07	-.3921E+08	-.4437E+08	-.9049E+07	.1696E+07	.5260E+07

INTERNAL RATE OF RETURN = 9.18 PERCENT

PAYBACK PERIOD - FROM FIRST REVENUE YEAR
UNDISCOUNTED VALUE 5.63 YEARS
DISCOUNT RATE 0 PERCENT 9.14 YEARS
DISCOUNT RATE 5.0 PERCENT 13.61 YEARS
DISCOUNT RATE 10.0 PERCENT EXCESS 27 YEARS
DISCOUNT RATE 15.0 PERCENT EXCESS 27 YEARS

GOVERNMENT TAX INCOME NET OF BONUSES ETC.	YEAR 84	YEAR 85	YEAR 86	YEAR 87	YEAR 88	YEAR 89	YEAR 90	YEAR 91
UNDISCOUNTED INCOME	0	0	0	0	0	0	0	0
DISCOUNT RATE 0 PERCENT	0	0	0	0	0	0	0	0
DISCOUNT RATE 5.0 PERCENT	0	0	0	0	0	0	0	0
DISCOUNT RATE 10.0 PERCENT	0	0	0	0	0	0	0	0
DISCOUNT RATE 15.0 PERCENT	0	0	0	0	0	0	0	0

MECON (MINERAL EVALUATION SYSTEM VERSION 1.1 - 81/4/30) DATE 83/06/23, PAGE 3

ECONOMIC MODEL OF ANTARTICA PLATINUM METALS MINE

PD/PT=0.42 REAL PT PRICE RISE 2PC 10 PC INFLATN****NOTE .1000E+04 = 1000.0****

		YEAR 92	YEAR 93	YEAR 94	YEAR 95	YEAR 96	YEAR 97	YEAR 98	YEAR 99
*C*PT PRICE 1983 US D/TROY OZ		.5677E+03	.5790E+03	.5906E+03	.6024E+03	.6145E+03	.6268E+03	.6393E+03	.6521E+03
*C*PD PRICE 1983 US D/TROY OZ		.1554E+03	.1585E+03	.1616E+03	.1649E+03	.1682E+03	.1715E+03	.1750E+03	.1785E+03
*C*U/G PRODUCTION TONNES/YEAR		.7189E+06	.7462E+06	.7462E+06	.7462E+06	.7462E+06	.7462E+06	.7462E+06	.7462E+06
-C-U/G RESERVES TONNES (YR ST)		.1405E+08	.1346E+08	.1285E+08	.1224E+08	.1163E+08	.1102E+08	.1041E+08	.9803E+07
-C-U/G ROM GRADE GRAM PT/TONNE		.3941E+01	.3941E+01	.3941E+01	.3941E+01	.3941E+01	.3941E+01	.3941E+01	.3941E+01
TOTAL ANNUAL SALES REVENUE		.1263E+09	.1468E+09	.1644E+09	.1840E+09	.2061E+09	.2307E+09	.2584E+09	.2893E+09
***ANNUAL PERSONNEL COSTS	D	.2742E+08	.3058E+08	.3364E+08	.3701E+08	.4071E+08	.4478E+08	.4926E+08	.5418E+08
***ANNUAL SUPPLIES COSTS	D	.1114E+08	.1257E+08	.1383E+08	.1521E+08	.1673E+08	.1841E+08	.2025E+08	.2227E+08
***ANNUAL POWER COSTS	D	.5613E+07	.6338E+07	.6972E+07	.7669E+07	.8436E+07	.9280E+07	.1021E+08	.1123E+08
***ANNUAL TRANSPORT COSTS	D	.3319E+07	.3790E+07	.4169E+07	.4586E+07	.5045E+07	.5549E+07	.6104E+07	.6714E+07
***ANNUAL GEN + ADMIN COSTS	D	.2094E+08	.2304E+08	.2534E+08	.2788E+08	.3066E+08	.3373E+08	.3710E+08	.4081E+08
TOTAL ANNUAL OPERATING COSTS		.6843E+08	.7632E+08	.8395E+08	.9235E+08	.1016E+09	.1117E+09	.1229E+09	.1352E+09
***EXPLORATION CAPEX	D	0	0	0	0	0	0	0	0
***SHAFT AND EQUIPMENT CAPEX	D	0	0	0	0	0	0	0	0
***UNDERGROUND MINING CAPEX	D	0	0	0	0	0	0	0	0
***TREATMENT PLANT CAPEX	D	0	0	0	0	0	0	0	0
***GENERAL FACILITIES CAPEX	D	0	0	0	0	0	0	0	0
*** PROJECT OVERHEAD COSTS	D	0	0	0.	0	0	0	0	0
***WORKING CAPITAL	D	.3278E+07	.2630E+07	.2544E+07	.2798E+07	.3078E+07	.3386E+07	.3725E+07	.4097E+07
TOTAL ANNUAL CAPITAL COSTS		.3278E+07	.2630E+07	.2544E+07	.2798E+07	.3078E+07	.3386E+07	.3725E+07	.4097E+07
ANNUAL RECOVERED CAPITAL COSTS		0	0	0	0	0	0	0	0
UNADJUSTED CAPITAL ALLOWANCES		0	0	0	0	0	0	0	0
ANNUAL TAXATION ALLOWANCES		.5791E+08	.7050E+08	.8042E+08	.2837E+08	0	0	0	0
ANNUAL TAXATION AT WBK RATE		0	0	0	.2216E+08	.3657E+08	.4165E+08	.4741E+08	.5394E+08
ANNUAL CASH FLOW		.5463E+08	.6787E+08	.7788E+08	.8890E+08	.7924E+08	.7904E+08	.9008E+08	.1026E+09
CUMMULATIVE CASH FLOW		-.2021E+09	-.1342E+09	-.5636E+08	.3254E+08	.1118E+09	.1908E+09	.2809E+09	.3835E+09
PRESENT VALUES AT SELECTED DISCOUNT RATES									
DISCOUNT RATE 0 PERCENT		.2317E+08	.2617E+08	.2730E+08	.2832E+08	.2295E+08	.2081E+08	.2156E+08	.2233E+08
DISCOUNT RATE 5.0 PERCENT		.1494E+08	.1606E+08	.1596E+08	.1577E+08	.1217E+08	.1051E+08	.1037E+08	.1023E+08
DISCOUNT RATE 10.0 PERCENT		.9926E+07	.1009E+08	.9567E+07	.9025E+07	.6649E+07	.5481E+07	.5162E+07	.4860E+07
DISCOUNT RATE 15.0 PERCENT		.6586E+07	.6468E+07	.5867E+07	.5294E+07	.3731E+07	.2942E+07	.2650E+07	.2386E+07

INTERNAL RATE OF RETURN = 9.18 PERCENT

PAYBACK PERIOD - FROM FIRST REVENUE YEAR
UNDISCOUNTED VALUE 5.63 YEARS
DISCOUNT RATE 0 PERCENT 9.14 YEARS
DISCOUNT RATE 5.0 PERCENT 13.61 YEARS
DISCOUNT RATE 10.0 PERCENT EXCESS 27 YEARS
DISCOUNT RATE 15.0 PERCENT EXCESS 27 YEARS

GOVERNMENT TAX INCOME NET OF BONUSES ETC.	YEAR 92	YEAR 93	YEAR 94	YEAR 95	YEAR 96	YEAR 97	YEAR 98	YEAR 99
UNDISCOUNTED INCOME	0	0	0	0	.2216E+08	.3657E+08	.4165E+08	.4741E+08
DISCOUNT RATE 0 PERCENT	0	0	0	0	.6420E+07	.9630E+07	.9970E+07	.1032E+08
DISCOUNT RATE 5.0 PERCENT	0	0	0	0	.3405E+07	.4864E+07	.4796E+07	.4727E+07
DISCOUNT RATE 10.0 PERCENT	0	0	0	0	.1860E+07	.2536E+07	.2387E+07	.2245E+07
DISCOUNT RATE 15.0 PERCENT	0	0	0	0	.1043E+07	.1361E+07	.1225E+07	.1103E+07

MECON (MINERAL EVALUATION SYSTEM VERSION 1.1 - 81/4/30) DATE 83/06/23, PAGE 4

ECONOMIC MODEL OF ANTARTICA PLATINUM METALS MINE

PD/PT=0.42 REAL PT PRICE RISE 2PC 10 PC INFLATN****NOTE .1000E+04 = 1000.0****

		YEAR 00	YEAR 01	YEAR 02	YEAR 03	YEAR 04	YEAR 05	YEAR 06	YEAR 07
*C*PT PRICE 1983 US D/TROY OZ		.6651E+03	.6784E+03	.6920E+03	.7058E+03	.7199E+03	.7343E+03	.7490E+03	.7640E+03
*C*PD PRICE 1983 US D/TROY OZ		.1820E+03	.1857E+03	.1894E+03	.1932E+03	.1970E+03	.2010E+03	.2050E+03	.2091E+03
*C*U/G PRODUCTION TONNES/YEAR		.7462E+06	.7462E+06	.7462E+06	.7462E+06	.7462E+06	.7462E+06	.7462E+06	.7462E+06
-C-U/G RESERVES TONNES (YR ST)		.9192E+07	.8582E+07	.7972E+07	.7362E+07	.6752E+07	.6142E+07	.5532E+07	.4922E+07
-C-U/G ROM GRADE GRAM PT/TONNE		.3941E+01	.3941E+01	.3941E+01	.3941E+01	.3941E+01	.3941E+01	.3941E+01	.3941E+01
TOTAL ANNUAL SALES REVENUE		.3240E+09	.3628E+09	.4063E+09	.4551E+09	.5097E+09	.5709E+09	.6394E+09	.7162E+09
***ANNUAL PERSONNEL COSTS	D	.5960E+08	.6556E+08	.7212E+08	.7933E+08	.8726E+08	.9599E+08	.1056E+09	.1161E+09
***ANNUAL SUPPLIES COSTS	D	.2450E+08	.2695E+08	.2964E+08	.3261E+08	.3587E+08	.3946E+08	.4340E+08	.4774E+08
***ANNUAL POWER COSTS	D	.1235E+08	.1359E+08	.1494E+08	.1644E+08	.1808E+08	.1989E+08	.2188E+08	.2407E+08
***ANNUAL TRANSPORT COSTS	D	.7386E+07	.8124E+07	.8937E+07	.9831E+07	.1081E+08	.1190E+08	.1308E+08	.1439E+08
***ANNUAL GEN + ADMIN COSTS	D	.4489E+08	.4938E+08	.5432E+08	.5975E+08	.6573E+08	.7230E+08	.7953E+08	.8748E+08
TOTAL ANNUAL OPERATING COSTS		.1487E+09	.1636E+09	.1800E+09	.1980E+09	.2178E+09	.2395E+09	.2635E+09	.2898E+09
***EXPLORATION CAPEX	D	0	0	0	0	0	0	0	0
***SHAFT AND EQUIPMENT CAPEX	D	0	0	0	0	0	0	0	0
***UNDERGROUND MINING CAPEX	D	0	0	0	0	0	0	0	0
***TREATMENT PLANT CAPEX	D	0	0	0	0	0	0	0	0
***GENERAL FACILITIES CAPEX	D	0	0	0	0	0	0	0	0
*** PROJECT OVERHEAD COSTS	D	0	0	0	0	0	0	0	0
***WORKING CAPITAL	D	.4507E+07	.4958E+07	.5453E+07	.5999E+07	.6599E+07	.7259E+07	.7984E+07	.8783E+07
TOTAL ANNUAL CAPITAL COSTS		.4507E+07	.4958E+07	.5453E+07	.5999E+07	.6599E+07	.7259E+07	.7984E+07	.8783E+07
ANNUAL RECOVERED CAPITAL COSTS		0	0	0	0	0	0	0	0
UNADJUSTED CAPITAL ALLOWANCES		0	0	0	0	0	0	0	0
ANNUAL TAXATION ALLOWANCES		0	0	0	0	0	0	0	0
ANNUAL TAXATION AT WBK RATE		.6134E+08	.6973E+08	.7923E+08	.8999E+08	.1022E+09	.1160E+09	.1316E+09	.1492E+09
ANNUAL CASH FLOW		.1168E+09	.1329E+09	.1512E+09	.1719E+09	.1953E+09	.2219E+09	.2520E+09	.2860E+09
CUMMULATIVE CASH FLOW		.5003E+09	.6333E+09	.7844E+09	.9563E+09	.1152E+10	.1374E+10	.1626E+10	.1912E+10
PRESENT VALUES AT SELECTED DISCOUNT RATES									
DISCOUNT RATE 0 PERCENT		.2311E+08	.2391E+08	.2472E+08	.2555E+08	.2640E+08	.2726E+08	.2814E+08	.2904E+08
DISCOUNT RATE 5.0 PERCENT		.1008E+08	.9934E+07	.9783E+07	.9630E+07	.9475E+07	.9318E+07	.9161E+07	.9003E+07
DISCOUNT RATE 10.0 PERCENT		.4573E+07	.4300E+07	.4042E+07	.3798E+07	.3567E+07	.3349E+07	.3142E+07	.2948E+07
DISCOUNT RATE 15.0 PERCENT		.2148E+07	.1932E+07	.1737E+07	.1561E+07	.1402E+07	.1259E+07	.1130E+07	.1014E+07

INTERNAL RATE OF RETURN = 9.18 PERCENT

PAYBACK PERIOD - FROM FIRST REVENUE YEAR
UNDISCOUNTED VALUE 5.63 YEARS
DISCOUNT RATE 0 PERCENT 9.14 YEARS
DISCOUNT RATE 5.0 PERCENT 13.61 YEARS
DISCOUNT RATE 10.0 PERCENT EXCESS 27 YEARS
DISCOUNT RATE 15.0 PERCENT EXCESS 27 YEARS

GOVERNMENT TAX INCOME NET OF BONUSES ETC.	YEAR 00	YEAR 01	YEAR 02	YEAR 03	YEAR 04	YEAR 05	YEAR 06	YEAR 07
UNDISCOUNTED INCOME	.5394E+08	.6134E+08	.6973E+08	.7923E+08	.8999E+08	.1022E+09	.1160E+09	.1316E+09
DISCOUNT RATE 0 PERCENT	.1067E+08	.1103E+08	.1140E+08	.1178E+08	.1216E+08	.1255E+08	.1295E+08	.1336E+08
DISCOUNT RATE 5.0 PERCENT	.4656E+07	.4584E+07	.4512E+07	.4439E+07	.4365E+07	.4291E+07	.4216E+07	.4142E+07
DISCOUNT RATE 10.0 PERCENT	.2111E+07	.1984E+07	.1864E+07	.1751E+07	.1643E+07	.1542E+07	.1446E+07	.1356E+07
DISCOUNT RATE 15.0 PERCENT	.9917E+06	.8915E+06	.8011E+06	.7196E+06	.6461E+06	.5799E+06	.5203E+06	.4666E+06

MECON (MINERAL EVALUATION SYSTEM VERSION 1.1 - 81/4/30) DATE 83/06/23, PAGE 5

ECONOMIC MODEL OF ANTARTICA PLATINUM METALS MINE

PD/PT=0.42 REAL PT PRICE RISE 2PC 10 PC INFLATN****NOTE .1000E+04 = 1000.0****

		YEAR 08	YEAR 09	YEAR 10	YEAR 11	YEAR 12	YEAR 13	YEAR 14	YEAR 15
*C*PT PRICE 1983 US D/TROY OZ		.7793E+03	.7949E+03	.8108E+03	.8270E+03	.8435E+03	.8604E+03	.8776E+03	.8952E+03
*C*PD PRICE 1983 US D/TROY OZ		.2133E+03	.2175E+03	.2219E+03	.2263E+03	.2309E+03	.2355E+03	.2402E+03	.2450E+03
*C*U/G PRODUCTION TONNES/YEAR		.7462E+06	.7462E+06	.7462E+06	.7462E+06	.7462E+06	.7462E+06	.7462E+06	.5074E+05
-C-U/G RESERVES TONNES (YR ST)		.4312E+07	.3702E+07	.3092E+07	.2482E+07	.1872E+07	.1262E+07	.6515E+06	.4148E+05
-C-U/G ROM GRADE GRAM PT/TONNE		.3941E+01	.3941E+01	.3941E+01	.3941E+01	.3941E+01	.3941E+01	.3941E+01	.3941E+01
TOTAL ANNUAL SALES REVENUE		.8022E+09	.8986E+09	.1007E+10	.1128E+10	.1263E+10	.1415E+10	.1585E+10	.1208E+09
***ANNUAL PERSONNEL COSTS	D	.1278E+09	.1405E+09	.1546E+09	.1700E+09	.1871E+09	.2058E+09	.2263E+09	.1693E+08
***ANNUAL SUPPLIES COSTS	D	.5252E+08	.5777E+08	.6355E+08	.6990E+08	.7689E+08	.8458E+08	.9304E+08	.6959E+07
***ANNUAL POWER COSTS	D	.2648E+08	.2912E+08	.3204E+08	.3524E+08	.3876E+08	.4264E+08	.4690E+08	.3508E+07
***ANNUAL TRANSPORT COSTS	D	.1583E+08	.1742E+08	.1916E+08	.2107E+08	.2318E+08	.2550E+08	.2805E+08	.2098E+07
***ANNUAL GEN + ADMIN COSTS	D	.9623E+08	.1059E+09	.1164E+09	.1281E+09	.1409E+09	.1550E+09	.1705E+09	.1275E+08
TOTAL ANNUAL OPERATING COSTS		.3188E+09	.3507E+09	.3858E+09	.4243E+09	.4668E+09	.5135E+09	.5648E+09	.4224E+08
***EXPLORATION CAPEX	D	0	0	0	0	0	0	0	0
***SHAFT AND EQUIPMENT CAPEX	D	0	0	0	0	0	0	0	0
***UNDERGROUND MINING CAPEX	D	0	0	0	0	0	0	0	0
***TREATMENT PLANT CAPEX	D	0	0	0	0	0	0	0	0
***GENERAL FACILITIES CAPEX	D	0	C	0	0	0	0	0	0
*** PROJECT OVERHEAD COSTS	D	0	0	0	0	0	0	0	0
***WORKING CAPITAL	D	.9661E+07	.1063E+08	.1169E+08	.1286E+08	.1414E+08	.1556E+08	.1712E+08	0
TOTAL ANNUAL CAPITAL COSTS		.9661E+07	.1063E+08	.1169E+08	.1286E+08	.1414E+08	.1556E+08	.1712E+08	0
ANNUAL RECOVERED CAPITAL COSTS		0	0	0	0	0	0	0	.1742E+09
UNADJUSTED CAPITAL ALLOWANCES		0	0	0	0	0	0	0	0
ANNUAL TAXATION ALLOWANCES		0	0	0	0	0	0	0	0
ANNUAL TAXATION AT WBK RATE		.1692E+09	.1918E+09	.2173E+09	.2461E+09	.2788E+09	.3156E+09	.3572E+09	.2749E+08
ANNUAL CASH FLOW		.3245E+09	.3681E+09	.4174E+09	.4731E+09	.5361E+09	.6074E+09	.6879E+09	-.1045E+09
CUMMULATIVE CASH FLOW		.2236E+10	.2604E+10	.3021E+10	.3495E+10	.4031E+10	.4638E+10	.5326E+10	.5222E+10
PRESENT VALUES AT SELECTED DISCOUNT RATES									
DISCOUNT RATE 0 PERCENT		.2995E+08	.3088E+08	.3184E+08	.3281E+08	.3380E+08	.3481E+08	.3584E+08	-.4950E+07
DISCOUNT RATE 5.0 PERCENT		.8844E+07	.8686E+07	.8527E+07	.8369E+07	.8211E+07	.8054E+07	.7898E+07	-.1039E+07
DISCOUNT RATE 10.0 PERCENT		.2764E+07	.2591E+07	.2428E+07	.2275E+07	.2131E+07	.1995E+07	.1867E+07	-.2344E+06
DISCOUNT RATE 15.0 PERCENT		.9098E+06	.8158E+06	.7313E+06	.6553E+06	.5870E+06	.5257E+06	.4707E+06	-.5652E+05

INTERNAL RATE OF RETURN = 9.18 PERCENT

PAYBACK PERIOD - FROM FIRST REVENUE YEAR
UNDISCOUNTED VALUE 5.63 YEARS
DISCOUNT RATE 0 PERCENT 9.14 YEARS
DISCOUNT RATE 5.0 PERCENT 13.61 YEARS
DISCOUNT RATE 10.0 PERCENT EXCESS 27 YEARS
DISCOUNT RATE 15.0 PERCENT EXCESS 27 YEARS

GOVERNMENT TAX INCOME NET OF BONUSES ETC.	YEAR 08	YEAR 09	YEAR 10	YEAR 11	YEAR 12	YEAR 13	YEAR 14	YEAR 15
UNDISCOUNTED INCOME	.1492E+09	.1692E+09	.1918E+09	.2173E+09	.2461E+09	.2788E+09	.3156E+09	.3572E+09
DISCOUNT RATE 0 PERCENT	.1377E+08	.1420E+08	.1463E+08	.1507E+08	.1552E+08	.1597E+08	.1644E+08	.1692E+08
DISCOUNT RATE 5.0 PERCENT	.4067E+07	.3992E+07	.3918E+07	.3844E+07	.3770E+07	.3696E+07	.3623E+07	.3551E+07
DISCOUNT RATE 10.0 PERCENT	.1271E+07	.1191E+07	.1116E+07	.1045E+07	.9782E+06	.9155E+06	.8566E+06	.8013E+06
DISCOUNT RATE 15.0 PERCENT	.4184E+06	.3750E+06	.3360E+06	.3010E+06	.2695E+06	.2413E+06	.2159E+06	.1932E+06

MECON (MINERAL EVALUATION SYSTEM VERSION 1.1 - 81/4/30) DATE 83/06/23, PAGE 6

ECONOMIC MODEL OF ANTARTICA PLATINUM METALS MINE

PD/PT=0.42 REAL PT PRICE RISE 2PC 10 PC INFLATN****NOTE .1000E+04 = 1000.0****

		YEAR 16	TOTAL
*C*PT PRICE 1983 US D/TROY OZ		.9131E+03	
*C*PD PRICE 1983 US D/TROY OZ		.2499E+03	
*C*U/G PRODUCTION TONNES/YEAR		0	
-C-U/G RESERVES TONNES (YR ST)		0	
-C-U/G ROM GRADE GRAM PT/TONNE		0	
TOTAL ANNUAL SALES REVENUE		0	.1398E+11
***ANNUAL PERSONNEL COSTS	D	0	.2272E+10
***ANNUAL SUPPLIES COSTS	D	0	.9325E+09
***ANNUAL POWER COSTS	D	0	.4701E+09
***ANNUAL TRANSPORT COSTS	D	0	.2807E+09
***ANNUAL GEN + ADMIN COSTS	D	0	.1715E+10
TOTAL ANNUAL OPERATING COSTS		0	.5671E+10
***EXPLORATION CAPEX	D	0	.9841E+07
***SHAFT AND EQUIPMENT CAPEX	D	0	.2795E+08
***UNDERGROUND MINING CAPEX	D	0	.4791E+08
***TREATMENT PLANT CAPEX	D	0	.1156E+09
***GENERAL FACILITIES CAPEX	D	0	.7959E+08
*** PROJECT OVERHEAD COSTS	D	0	.1916E+08
***WORKING CAPITAL	D	0	.1883E+09
TOTAL ANNUAL CAPITAL COSTS		0	.4883E+09
ANNUAL RECOVERED CAPITAL COSTS		.1408E+08	.1883E+09
UNADJUSTED CAPITAL ALLOWANCES		0	.3000E+09
ANNUAL TAXATION ALLOWANCES		0	.3000E+09
ANNUAL TAXATION AT WBK RATE		0	.2804E+10
ANNUAL CASH FLOW		-.1340E+08	.5208E+10
CUMMULATIVE CASH FLOW		.5208E+10	.5208E+10
PRESENT VALUES AT SELECTED DISCOUNT RATES			
DISCOUNT RATE 0 PERCENT		-.5771E+06	.4517E+09
DISCOUNT RATE 5.0 PERCENT		-.1154E+06	.9814E+08
DISCOUNT RATE 10.0 PERCENT		-.2485E+05	-.1064E+08
DISCOUNT RATE 15.0 PERCENT		-.5731E+04	-.4441E+08
INTERNAL RATE OF RETURN = 9.18 PERCENT			
PAYBACK PERIOD - FROM FIRST REVENUE YEAR			
UNDISCOUNTED VALUE		5.63 YEARS	
DISCOUNT RATE 0 PERCENT		9.14 YEARS	
DISCOUNT RATE 5.0 PERCENT		13.61 YEARS	
DISCOUNT RATE 10.0 PERCENT		EXCESS 27 YEARS	
DISCOUNT RATE 15.0 PERCENT		EXCESS 27 YEARS	
GOVERNMENT TAX INCOME NET OF BONUSES ETC.			
UNDISCOUNTED INCOME		.2749E+08	.2804E+10
DISCOUNT RATE 0 PERCENT		.1183E+07	.2559E+09
DISCOUNT RATE 5.0 PERCENT		.2365E+06	.8369E+08
DISCOUNT RATE 10.0 PERCENT		.5096E+05	.3095E+08
DISCOUNT RATE 15.0 PERCENT		.1175E+05	.1271E+08

ANTARCTICA—SELECTED ANNOTATED BIBLIOGRAPHY
(modified and expanded after Kimball 1983)

GENERAL

Current Antarctic literature and the Antarctic bibliography (1951–1984). Prepared by the Cold Regions Bibliography Project, Science and Technology Division, Library of Congress, for the Division of Polar Programs National Science Foundation, Vol. 1–13. Government Printing Office, Washington, DC 20402.

This bibliography comprises more than 31 000 citations and abstracts and is without doubt the most comprehensive overview of the world's Antarctic literature. It contains abstracts of all serious Antarctic work published, with a main emphasis on scientific research contributions. Abstracts of up to approximately 200 words are presented under 13 subject categories: (a) general; (b) biological sciences; (c) cartography; (d) expeditions; (e) geological sciences; (f) ice and snow; (g) logistics, equipment, and supplies; (h) medical sciences; (i) meteorology; (j) oceanography; (k) atmospheric physics; (l) terrestrial physics; (m) political geography. Citations and abstracts are cumulated every 13 months into hardcover volumes, each of which contains author, subject, geographic, and grantee indexes.

James N. Barnes (1982). *Let's save Antarctica* (Photography by Eliot Porter). Greenhouse Publications, PTY Ltd. Richmond, Australia (28 pp. text, 50 pp appendices). Copies available from The Antarctica Project, 624 9th Street NW, 5th Floor, Washington, DC 20001.

The author's passionate belief in preserving Antarctica's natural beauties—illustrated most convincingly by the Porter photographs—and its values as a scientific reserve is combined with a practical 'how-to' manual for those he would recruit to the cause. He presents background on the Antarctic Treaty and the resources of Antarctica and sets forth in a series of appendices critical terms, acronyms, a bibliography, copies of relevant Treaty documents and copies of letters and resolutions for those most concerned with saving Antarctica from environmental devastation. The text contains three recurring themes: (a) that governments must demonstrate an increased commitment to fund and conduct the research required to understand and protect Antarctic ecosystems and to implement the legal regimes agreed upon; (b) that Antarctic ecosystems could be significantly disrupted by ill-controlled resources development; and (c) that Antarctic Treaty meetings are too secretive and Treaty Parties too short-sighted to ensure adequate protection of Antarctica.

Barney Brewster (1982). *Antarctica: wilderness at risk, Friends of the Earth, New Zealand and Australia 125 pp.*

This book describes Antarctica and presents an informative, narrative account of man's impact on the Southern Continent. Pictures and maps complement the text throughout. The author reviews the history and politics of national activities in Antarctica as well as current international science there. He then covers resource surveys in Antarctica, the potential hazards of development activities, and the interplay of resources and politics under the Antarctic Treaty system. This book concentrates on the involvement of New Zealand and Australia in these areas.

In a concluding chapter the author assesses the minerals politics of Antarctica. He underscores what could be the pivotal role of the Scientific Council for Antarctic Research (SCAR), Committee of the International Council of Scientific Unions (ICSU), the only international force in the area other than the Antarctic Treaty. While its members include those most conversant with Antarctic science and technology and the hazards of human activity there, he regrets that they are constrained by SCAR's apolitical charter. Finally, he dismisses as unrealistic environmentalists' efforts to internationalize Antarctica as a world park under United Nations auspices. He argues instead for national initiatives to place Antarctic territories under the United Nations Convention for the Protection of World Cultural and Natural Heritage, as advocated by Friends of the Earth (FOE) Australia.

Philip W. Quigg (1983). *A pole apart: the emerging issue of Antarctica.* A Twentieth Century Fund report, New Press, McGraw-Hill Book Company, 299 pp.

In this book, Quigg projects a political kaleidoscope against the background of Antarctica's historical exploration and exploitation, legal status, scientific findings, socio-economic facts, and political stakes. A monumental undertaking that this author achieves to a remarkable degree. Although no scholarly work in depth, this book serves as an introductory manual for those interested in the international affairs of Antarctica and the dubious policies of its 14 'guardian' nations with their self-formulated rules and regulations condensed in the Antarctic Treaty and various conventions. The book benefits from Quigg's capability to assess and overview many future political scenarios in the ongoing 'tug of war' over Antarctica, and specifically the role played herein by successive US administrations.

Clearly, beneath the guardians' cover of quibble and quarrels boils a battle for Antarctica's natural resources. Unfortunately Quigg's scientific assessment of these potential resources is the weakest link in this book. But this is not entirely his fault; Antarctic scientists themselves must be largely to blame. Whereas Antarctica has up to now been accepted as communal scientific ground, it is in danger of becoming a 'scientific colony'. Scientists are not delivering the factual goods on which vital social, economic, and political decisions of global significance must be made before too long. Quigg's candour throughout this book serves as a warning for the scientific community at large. We should be grateful for his signals, and for the benefit of all, we ought to make this book compulsory reading for undergraduate and graduate science students in an effort towards awakening them to their social responsibilities as scientists and the potential consequences of their future work.

Barbara Mitchell and Jon Tinker (1980). *Antarctica and its resources* Earthscan, London 98 pp. Available from the International Institute for Environment and Development (IIED), 10 Percy Street, London W1P 0DR, UK or 1319 F Street, NW Suite 800, Washington DC 20004, USA.

The background chapters answer the following questions: What is Antarctica? Who discovered it? Who claims it? What is the Antarctic Treaty? What is SCAR? Why are scientists interested? Who lives there? The second section on the resources of Antarctica looks at Antarctic geology and the Gondwanaland supercontinent hypothesis: Is there any oil? Are there minerals on land? What would be the environmental impacts of exploitation? A section on krill explains which nations have been interested in krill and why, how much krill may be caught, the risks of overfishing and the technological problems of krill recovery and processing. The remaining chapters in this section look at other living resources such as finfish and whales as well as at ice and tourism. The final section examines Antarctic politics: Should there be a ban on minerals development? It looks at the politics of krill and fish with comment on the emerging Convention on the Conservation of Antarctic Marine Living Resources. Can the Treaty cope with sovereignty? What is the conflict between the international community and the Club? The document concludes by looking at national policies and analysing the choices.

This booklet was originally published by Earthscan as a briefing document for the international press and is therefore not set out in a conventional format. It is designed for rapid reference and assimilation of facts.

L. Kimball (1983*b*). *Antarctica: a continent in transition.* A portfolio prepared by the International Institute for Environment and Development (IIED), London.

This handy fold-out portfolio, measuring 22.5 cm by 30.5 cm, serves as a sound general introduction to Antarctic affairs. It contains 11 looseleaved fact sheets each dealing with one of the following topics: (i) Antarctic vital statistics; (ii) history and claims; (ii) living resources; (iv) mineral resources; (v) resources exploitation; (vi) national activities in Antarctica; (vii) the Antarctic Treaty system; (viii) the Antarctic Treaty; (ix) the Convention for the Conservation of Antarctic seals; (x) the Convention on the Conservation of Antarctic Marine Living

Resources; (xi) The Scientific Committee on Antarctic Research (SCAR) of the International Council of Scientific Unions (ICSU).

The fact sheets are laid out for fast 'executive' reading. Other information is treated under the following headings, each also on separate sheets; Treaty documents (including the text of the Antarctic Treaty; the Convention for the Conservation of Antarctic Seals; the Convention on the Conservation of Antarctic Marine Living Resources; agreed measures for the conservation of Antarctic fauna and flora, and Recommendations XI-I, Antarctic mineral resources); bibliography; maps showing Antarctic scientific stations and claims in Antarctica. Also included in the folio are four topical articles and two opinion pieces on Antarctic resources, politics, and environmental problems. For the uninitiated as well as the seasoned Antarctic specialist, this folio is an extremely informative and up-to-date overview.

EARTH SCIENCES AND MINERALS

John C. Behrendt, ed. (1983). *Petroleum and mineral resources of Antarctica* US Geological Survey Circular No. 909, 75 pp.

This collection of three detailed technical articles presents an overview of geological, geochemical, and geophysical data and research in Antarctica and its bearing on potential petroleum resources there; of known mineral occurrences, and of metals in the Antarctic Dufek intrusion. The papers highlight the need for a great deal more research in order to assess effectively the mineral resources potential of Antarctica. They also note in passing the physical and political difficulties of minerals development at the South Pole and how this effects the economics of possible Antarctic minerals activities. The article on potential petroleum resource provides a useful summary of recent seismic reflection surveys of the Antarctic continental margins. Arthur Ford, the world's most knowledgeable geologist on the Dufek intrusion, gives a lucid overview of the known facts related to this immense geological anomaly.

M. W. Holdgate and J. Tinker (1979). *Oil and other minerals in the Antarctic;* the environmental implications of possible mineral exploration or exploitation in Antarctica.

Report of an international workshop of interested scientists sponsored by the Rockefeller Foundation in Bellagio, Italy, and known as the Bellagio Report.

Scientific Committee on Antarctic Research (SCAR) House of Print, London 51 pp. Out of print.

After describing the major geographical, political, and scientific features of the Antarctic, the report indicates that any proven minerals deposits onshore would have to be exceptionally rich to be economically exploitable, and that there is no immediate prospect of development. Offshore, the report states that Antarctica's continental margins may well contain petroleum, that exploration appears to present no insurmountable technological problems, and that methods of exploitation could probably be developed quickly if oil were discovered. The report written in 1979 predicts that rising energy prices might make exploration attractive in the near future and that exploitation might be a commercial proposition one day if a very large field were discovered.

With respect to the environmental effects of minerals development, the report indicates that not enough is known about Antarctic ecosystems to predict such effects in any detail. It notes that the scientific value of ecosystems on land and the potential economic importance of the growing krill fishery argue for caution. It calls for a selective, analytical approach to environmental impact assessment and research and monitoring programs tailored to the decision-making requirements of the ultimate management authority.

On the basis of a case study of possible oil development in the Ross Sea, the report identifies some specific environmental questions that a management authority would need to answer. It concludes that drilling in the exploration phase might result in oil spills of small-scale and uncertain effect on marine life. Once commercial exploitation began, a well blow-out or a major tanker accident might occur. But even a spill as large as 500 000 tons of crude oil would be unlikely to cause significant damage to the krill or seabird populations of Antarctica as a whole. On the other hand, repeated spills might have more serious, cummulative effects. Moreover, onshore oil facilities would have to be located in the limited ice-free areas of Antarctica, many of which are already used for scientific stations. In these areas, impacts, though local, might be considerable. The report also states that impacts of onshore minerals exploitation could be severe, even if of limited extent.

At its close the workshop recommended that SCAR initiate a study to investigate further the environmental impacts of Antarctic minerals development. The report argues that SCAR should take the lead in filling the research needs identified and notes that the credibility of Treaty nations will depend in part on their ability to develop appropriate environmental guidelines for minerals activities.

Campbell Craddock (ed. 1982) *Antarctic geoscience. Symposium on Antarctic geology and geophysics.* International Union of Geological Sciences, Series B, No. 4, 1172 pp. University of Wisconsin Press.

This massive volume clearly displays the great increase in output of data since IGY. It also illustrates the evolution of research objectives and methods, from the type linked with romanticized geological expeditions to a more mature scientific approach. *Antarctic Geoscience* is the result of international co-operations between earth scientists from at least 15 countries, and is the outcome of the 3rd Symposium of Antarctic Geology and Geophysics held in Wisconsin in 1977. It marks a milestone in the documentation of Antarctic data and Campbell Craddock, the main editor, has done the geoscience community an invaluable service, particularly those unfamiliar with Antarctica as a whole. This book brings together a wealth of data, which would otherwise be dispersed beyond recovery. Two hundred and twenty-eight authors have contributed 115 papers which are subdivided into 12 sections as follows; Gondwanaland (11 papers, constituting 9.2 per cent of the book's length). Scotia Arc Region (39 papers, 26.3 per cent), East Antarctic Shield (14 papers, 6.8 per cent), Upper Precambrian–Palaeozoic rocks (11 papers, 7.8 per cent), palaeontology (8 papers, 5.6 per cent), igneous rocks (14 papers, 8.6 per cent), structural geology and tectonics (9 papers, 5 per cent), mineral deposits (4 papers, 2.6 per cent), crustal structure (13 papers, 6 per cent), sub-glacial morphology (4 papers, 2.5 per cent, marine geology (8 papers, 4.3 per cent) and Cenozoic history (16 papers, 9.2 per cent). Inevitably in a book of this nature, there is much duplication of material, but the final index is sufficiently detailed to enable the reader to locate appropriate information.

The volume is clearly printed, liberal with quality maps, diagrams, photographs, tables in a text almost free of typographical errors, with standardized references attached to each paper. The references amount in total to almost 100 pages. There is an extremely useful index map of Antarctica on the inside cover, showing the areas discussed in each paper. Finally there is a copy of Craddock's own 1:5 000 000 geological map of Antarctica (published in 1972) in the back pocket.

R. Tingey, ed. (1985). *The Geology of Antarctica.* Oxford University Press, (forthcoming).

Despite almost 80 years of Antarctic geological research, no widely available overview of Antarctic geology has been published: geological data from this continent has, at best, been presented in symposium volumes, as compendiums of research papers. This forthcoming book offers, for the first time, a comprehensive, synoptic account of the geology of this vast continent; the last continent to be dealt with in this manner: long overdue. This book is tailored for a wide audience interested in Antarctica and its geological architecture as a whole, but also aims to attract the attention of specialists such as metamorphic and igneous petrologists, palaeontologists, stratigraphers, and others who, for so long, have 'had to disregard' data from this continent in their research and teaching efforts.

The book is a multi-author volume, but has been compiled and edited in such a way that it retains sufficient coherency between chapters and special topics. This has been possible because the editor, R. Tingey, is a geologist with a wide range of Antarctic Earth Science interests and experience, and because many of the individual contributors, workers who have devoted much of their careers to specific aspects of Antarctic geoscience, are well informed about each other's work. The lay-out of the book is constructed within a geochronologic framework and contains the following chapters: (1) Pre-Cambrian geology: (2) Palaeozoic geology; (3) Mesozoic geology; (4) Cenozoic geology; (5) Glaciology; (6) Special topics. This latter chapter includes a wide spectrum of subjects such as the igneous geology of the Dufek complex, palaeobotany, vertebrate and invertebrate palaeontology, the Antarctic continental shelf and crustal structure, Antarctic tectonics, as well as the resource potential of Antarctica.

The only major shortcoming of the book is reflected in its Anglophilic roots. The contributing scientists are from New Zealand, Australia, USA, and the United Kingdom. This does not fairly reflect contributions from other nationals to Antarctic geoscience. Nevertheless, this book promises to be a valuable working tool.

ECONOMICS, POLITICS, AND MINERALS

Francisco Errego Vicuña, ed. (1983). *Antarctic resources policy: scientific, legal and political issues.* Cambridge University Press, Cambridge, UK 335 pp.

This book represents the official record of the first International political conference held on the Antarctic continent, at the Chilean Antarctic scientific station, Teniente March, under the auspices of the Chilian Air Force, the Tinker Foundation and the Chilean Foreign Ministry. The 23 contributions which make up this book span the entire spectrum of debates which bear influence on the negotiations towards establishing a minerals regime for Antarctica. As an introduction, there is a useful overview of these Antarctic affairs by Vicuña, the editor, and Director of the Institute of International Studies of the University of Chile. This is followed by five sections, as follows:

1. The state of Antarctic knowledge and experience. This first section includes three scientific overviews concerning living, mineral (oil) and environmental resources.
2. The policy for the conservation of the living resources of Antarctic.
3. The policy for the exploration and exploitation of the mineral resources of Antarctica. This section includes 'an overview of the problems which should be addressed in the preparation of a regime governing the mineral resources of Antarctica', by Beeby of the 'Beeby Draft' which is presently believed to be the Antarctic Treaty nations' working model for a possible future Antarctic minerals regime.
4. Issues on Antarctica and the Law of the Sea.
5. The policy for Antarctic co-operation.

For any reader interested in Antarctica, this book provides a bird's eye view of its historical and contemporary affairs. The view is however almost exclusively that of nationals from countries who are Consultative nations to the Antarctic Treaty and almost half of the 21 contributors are government or ex-government officials of these nations. The sole exception is an overview of recent resource developments in the Arctic, with emphasis on technology, by Roots, a Canadian science advisor to the Department of the Environment, Canada. There are only a few thorough informative scientific papers: Gjelsvik (Director of the Norwegian Polar Research Institute) presents a sound scientific report on oil, in which he outlines *inter alia* the major sedimentary basins of Antarctica which might yield oil. Gjelsvik stresses however that nothing is known about their oil potential, their thermal history, and even less about any source and reservoir rocks. Knox (Professor of Zoology, University of Canterbury, New Zealand) provides a lucid scientific account of the Southern Oceans' living resources, their habitats and their interdependencies. He includes factual data which shows why the whale population can never get back to what it was, given the present (increasing) krill harvesting. Unfortunately the book does not provide an adequate geological account of Antarctica's (economic) mineral resources.

Most of the remainder of the book focuses on the socio-political questions of conservation as opposed to exploitation of Antarctica's exhaustible resources and the legal means with which to control and share the returns thereof and the accompanying responsibilities for Antarctic environmental protection. This collection of thoughts and facts provides the reader with some definite insights into the philosophical and moral attitudes (and commitments) of Antarctic Treaty nations. There are some good and thought-provoking articles by veteran Antarctic diplomats such as Brennan and Van der Essen (former ambassadors for Australia and Belgium, respectively). Basically however, the volume is a production by the boys (of the Treaty Nations) for the boys and any others who may wish to join 'the club' on its well-established terms. All these contributions reflect sentiments of a need to continue the *modus operandi* of the Antarctic Treaty, with very small concessions to the rest of the world. All are adament that an International body should be kept out of Antarctic political affairs and that Antarctica is not yet ready to be released from its present guardianship; the Antarctic Treaty 'parents' are determined to hang on to 'their baby' for some time to come.

Giulio Pontecorvo (1982). The economics of the resources of Antarctica. In *New nationalism and the use of common spaces,* (ed. Jonathan I. Charney) Allanheld, Osman & Co., Montclair, NJ. 11 pp.

The author examines data on the non-living resources of Antarctica and cost estimates of their production. He concludes that under market conditions these resources will not be exploited within 20 years and probably for a much longer period. Nevertheless, he raises the possibility that political motives, such as the desire to preserve one's ability to participate in potential future revenues, may encourage uneconomic exploitation. His final recommendation is that a fund be established to further public knowledge both of basic scientific information and of 'operational' information of the type required for exploitation—information on economic and engineering factors and on the environment. Money would be made available from a tax on living resources exploitation in Antarctica, supplemented by fees imposed on Treaty Parties and on others active or desiring to be active in Antarctica. He maintains that keeping this knowledge in the public domain, rather than as proprietary corporate or national data, will 'help avoid a national scramble for control of the unknown'.

Barbara Mitchell (1983). *Frozen stakes; the future of Antarctic minerals.* Earthscan, London 135 pp. Available from the IIED.

This monograph is based on an earlier report prepared for the US government. The author explores different options for the management of Antarctic minerals. It begins with a description of the resources at stake, the technological and economic aspects of exploitation and the environmental implications of mineral-related activities in Antarctica. It then turns to polar politics: the history of exploration and claims, the Antarctic Treaty, the Scientific Committee on Antarctic Research (SCAR) and scientific research, the interests of the wider international community, the implications of the Falklands/Malvinas conflict for Antarctica, living resources and their management, and the initial phase of the mineral resources negotiations.

The following requirements for a minerals regime are then identified and examined: acceptability to claimants, acceptability to non-claimants; aceptability to the Treaty Parties as members of a group; acceptability to the wider international community; acceptability to those wishing to explore and exploit; environmental protection and maintenance of demilitarization and scientific cooperation. In the light of these criteria, five possible regimes are evaluated. The first three are extreme solutions: a world park where exploration and exploitation would be barred; a global regime, where the equivalent of the International Seabed Authority contemplated in the 1982 Law of the Sea Convention would be established, and a condominium where Treaty parties would pool sovereignty. The fourth and fifth regimes represent compromise solutions: a system of diluted sovereign rights for claimants, and a system of jurisdictional ambiguity where territorial aspirations are met through side deals. The author concludes that the latter regime provides the best alternative and notes that the minerals negotiations have recently taken a turn in this direction. She calls attention to the weakness of this system in terms of international community interests and environmental protection and suggests ways of dealing with these problems.

James H. Zumberge (1982). Potential mineral resource availability and possible environmental problems in Antarctica. In: *New nationalism and the use of common spaces,* (ed. Johnathan I. Charney). Allenheld, Osman & Co. Montclair, NJ. 40 pp.

The author describes the continent of Antarctica, its geography and geology, the nature of the Southern Ocean, the climate of Antarctica and Antarctic life forms. He then explores Antarctica's mineral resource potential (based on the work of the Scientific Committee on Antarctic Research), including resources of the deep seabed and ice. He notes that mineral resources in Antarctica have no economic significance today or in the near future. The most detailed section of the article assesses the environmental risks associated with different types of minerals exploration and exploitation, drawing extensively on related scientific work in Antarctica, environmental monitoring there, and similar research and development experiences elsewhere. With respect to the possible impact of an oil spill of the order of 2.5 million barrels of oil in Antarctic waters (comparable to the IXTOC-1 spill in the Bay of Campeche off Mexico), he concludes that the variables involved make any definitive assessment pure speculation, with impacts ranging from negligible to serious. He affirms, however, that minerals exploration in Antarctica—whether on land or offshore—does not pose major risks to the environments. He does not believe that there exists a sound basis for decisions on offshore exploitation within the constraints imposed by environmental consideration, and he suggests research programmes to help fill this information gap.

Willian E. Westermeyer, Woods Hole Oceanographic Institution (1984). *The politics of mineral resource development in Antarctica: Alternative regimes for the future.* Westview press: Published in co-operation with the Institute for Marine and Coastal Studies, University of Southern California, 200 pp.

This book examines the international legal issues surrounding future exploitation of resources in the Antarctic region. The present international regime—the Antarctic Treaty—has operated smoothly for 21 years, says Westermeyer, but it has allowed participating nations to sidestep disagreements concerning sovereignty. At the same time the Treaty lacks adequate measures concerning future mineral exploration. Serious negotiations for an acceptable regime have only recently begun, leaving the crucial issue of sovereign rights unresolved. Following an analysis of the Treaty, Westermeyer discussed twelve proposals viewed as most likely alternatives for a new international regime for the Antarctic region. Employing a method known as Multi-Alternative Utility Analysis, he evaluates each alternative in terms of the relevant interests of the 14 nations involved in order to determine which of the proposals best serves collective interests. Incorporated in his evaluation are estimates of the importance of these interests for each state and an assessment of the probable effect each alternative would have in satisfying them. Also addressed are many important but yet unresolved issues, such as the criteria for the nature of participation in an Antarctic minerals regime, potential conflict among various development objectives, and the balance between environmental and developmental concerns.

THE ANTARCTIC TREATY SYSTEM

Francis M. Auburn (1982). *Antarctic law and politics.* Hurst & Co., London 316 pp.

This is the most detailed and comprehensive study to date which examines the legal framework for Antarctic activities by Treaty nations, stressing issues of international law and the potential inadequacies of present Treaty arrangements for dealing with resource disputes among Treaty Parties. It begins by reviewing sovereignty in Antarctica and its basis in discovery, annexation, occupation, state acts, and the sector theory and notification of scientific activities under the Antarctic Treaty. The problems of ice and polar circumstances in relation to sovereignty are also considered. The author then looks at national activities in Antarctica, emphasizing the South American sector, where sovereignty is still publicly displayed, the history of US involvement, and the vigorous expansion of the Soviet Antarctic program in the mid-fifties. He describes in detail the Antarctic Treaty system, its origins, content, political history and implications, particularly in relation to jurisdiction. Antarctic resources and environmental issues are also treated in some detail. The author concludes by cautioning that past performance is no guarantee of continuing achievement: the Treaty system has proved a success so far but it is not clear, he argues, that it will prove adequate to cope with resource disputes among the parties themselves.

Frank C. Alexander, Jr. (1978). A recommended approach to the Antarctic resource problem. *University of Miami Law Review* **33**, 417–23.

The author suggest that the Southern polar region is ripe for future development, but notes the legal and political vacuum there. He describes the history of Antarctica and examines the principles underlying sovereign claims. After explaining the risks of continuing the status quo in Antarctica, he looks at what is needed for a solution of the Antarctic resource problem, exploring the common interests of claimants, non-claimants, and non-Treaty states. This leads to an analysis of four potential solutions: an international solution, recognition of sector claims by Treaty parties, submission of sovereignty claims to the International Court of Justice, and joint Antarctic sovereignty among the Parties. The author then proposes his own approach: joint Antarctic resources jurisdiction. Under this scenario the Antarctic Treaty Parties would declare joint and exclusive jurisdiction over the continent and the continental shelf of Antarctica, leaving the Treaty intact. No mineral resource activities would be permitted for five years, after which period shelf areas designated as suitable for commercial activities would be divided into concession blocks. At that time all interested states in the world community could bid for development rights. Revenues generated from a tax on exploitation would accrue to both Treaty Parties and to a United Nations trust fund craeted to aid developing countries. The author concludes that a system of joint Antarctic resources jurisdiction should be acceptable to claimant and non-claimant parties alike and to a sufficient number of the remaining members of the international community to ensure its viability.

W. M. Bush (1982). *Antarctica and international law: a collection of inter-state and national documents,* Vol. 1–3. Oceans Publications, London.

The author has compiled a complete collection of basic national materials on the international legal issues of Antarctica from 17 countries. He also includes a separate section on Antarctic Treaty documents and other documents of a multilateral nature. In addition to the 14 Antarctic Treaty Consultative Parties of 1983, Peru and Uruguay are included, as is pre-war Germany. This is a very thorough contribution of immense value to historians and international lawyers. Bush has refrained from including his own interpretations and speculations based on this vast data base.

Jonathan I. Charney (1982). Future strategies for an Antarctic mineral resource regime—can the environment be protected? In: *New nationalism and the use of common spaces.* (ed. Jonathan I. Charney). Allanheld, Osman & Co., Montclair, NJ, 32 pp.

The author explores alternatives for an Antarctic mineral resources regime and their implications for policymakers, environmentalists, and commercial operators. He outlines the functions that would be covered by a regime, from data acquisition and standard-setting to enforcement and financial support. He then analyses three general approaches to accomodating these various interests and operational functions: (a) universal action by a comprehensive organisation open to all states; (b) unilateral action; and (c) a variety of limited multilateral approaches. He concludes that the third option holds most promise, particularly if a gradualist approach is applied to building the regime; that is, if the regime were initially restricted to resources exploration and scientific research.

David A. Colson (1980). The Antarctic treaty system: the mineral issue. *Law and Policy in International Business* **12**, (4), 62 pp.

The author underscores the effectiveness of the Antarctic Treaty consensus procedures as a means of obtaining agreement among Treaty Parties on issues and crcumstances not directly addressed in the 1959 Antarctic Treaty. He summarizes recommendations and other legal instruments which have pragmatically avoided the issue of national jurisdiction yet achieve common objectives. These cover environmental protection, resources conservation, non-interference with scientific investigation and problems related to increasing tourism. On this basis he makes a favourable assessment of the Treaty systems's capacity to meet the more recent challenge of possible Antarctic mineral resources development. He discusses the evolution of the minerals talks through 1979 in dealing with political, legal, and environmental considerations. His analysis focuses on the creativity of the Antarctic Treaty system in responding to issues which arise among the Parties; it does not evaluate the system responsiveness to challenge from outside.

Anonymous (1978). Thaw in international law? Rights in Antarctica under the law of common spaces. Note in *Yale Law Journal* **87**, 804–59.

The author argues that changing world needs and expectations require that the Antarctic and its non-renewable resources be governed by an emerging international law of common spaces. He refutes the theories invoked in support of claims in Antarctica and asserts that neither the Antarctic Treaty nor international practice substantiate exclusive regulation and control of Antarctica. In his conclusion he outlines an international regime for Antarctica based on the attributes of common spaces like the high seas, the deep seabed and outer space: they are unsuitable for extended human habitation; their principal benefits have not historically required exclusive control; they play a sensitive role in the Earth's ecology and they have emerged only in the post-colonial era as sites of critical value, thus disallowing national seizure and/or appropriation. He draws on existing common space regimes to identify seven principles for the Antarctic regime: peaceful use, common sovereignty, common rights of access, common rights to resources and in decision-making, and rights and obligations to protect the environment and to freely conduct scientific research and share the results.

Rainier Lagoni (1979). Antarctica's mineral resources in international law. *Zeitschrift für Auslandisches Offentliches Recht und Volkerecht, Sonderabdruck aus Bank* **29**, 37 pp.

The author examines the legal bases for title to Antarctic resources and for regulatory jurisdiction in light of the different views on claims in Antarctica. He concludes that the divergent positions of the Treaty Parties 'do not create a favourable economic climate for exploration and exploitation of its mineral resources'. He also raises questions about the legal status of the continental shelf off Antarctica and the rights of states not party to the Antarctic Treaty. The author then evaluates tentative legal and regulatory solutions to Antarctic mineral resources development: (1) to retain the status quo and allow states to exploit Antarctic resources unilaterally; (2) to nationalize Antarctica, either as individual states' territory or through a condominium arrangement among the Treaty parties; (3a) to internationalize fully the Antarctic continent, entailing a renunciation of the claims; (3b) to internationalize fully mineral resources in a manner which would leave sovereign control to the claimant states but provide for international rules and standards; (3c) to set up among Treaty Parties a limited system of international administration/regulation, or even development, which would reflect the interests of mankind as a whole through some form of revenue-sharing. He concludes that any Antarctic mineral resources regime must meet the following requirements: preserve peace and flexibility; support the purposes of the Antarctic Treaty; provide acceptable solutions to title and jurisdiction; protect operators from discrimination; provide for resolution of common problems such as environmental protection and safety; and take into account of the interests of all mankind through benefit-sharing.

M. C. W. Pinto (1978). The international community and Antarctica. *University of Miami Law Review* **33**, 475–87.

The author feels that the regulation and conservation of Antarctica must assume a position of major significance in the movement toward a new, equitable economic order, which would place greater emphasis on co-operation than on the pioneering, competitive spirit of an earlier time. He begins by arguing that a growing body of opinion holds that claims to sovereignty in Antarctica have no place in the world today. He maintains that Alexander's proposal (see above) to reconcile the interests of claimants, non-claimants, and the international community through the existing legal structure of the Antarctic Treaty may not appeal to non-Treaty states. Instead, he would utilize United Nations procedures to produce a comprehensive study of the economic potential of Antarctica, followed by a UN General Assembly resolution blueprinting an equitable solution. The resolution would neither address the status of Antarctica nor contain any implications for claims, but the Assembly would call for a Committee on Antarctica comprised of Treaty Parties and fifteen other countries. The Committee would recommend to the Assembly how best to reconcile the interests of the world community with the interests of individual countries, allowing Treaty Parties to propose an acceptable plan for environmental protection, resources exploitation, and scientific research. In the author's view, this approach would enable Treaty parties to remain influential but would involve them in an inevitable movement towards the new international economic order. He argues that consigning the problem of Antarctica to the gradualness of United Nations procedures affords an opportunity for the slow defusing of claimant sensibilities.

Deborah Shapley (1984). The seventh continent: Antarctica in a resource age. Resources for the Future, Washington DC.

The author examines the role of Antarctica in world affairs, past, present, and future. She also describes the background of US involvement in the region and present US policy options. The book is a guide for the policy analyst, explaining the web of international relationships that sprang up in the region earlier in this century, the geopolitical stakes of the US and the Soviet Union, and the circumstances of the negotiation of the Antarctic Treaty.

The author describes the extraordinary success of the Treaty and accompanying emphasis on science in keeping the region peaceful for the last 20 years. She emphasizes the emergence of resource issues, beginning in the late 1960s with the growth of fishing in Antarctic waters and the negotiations of the Convention to protect the Antarctic marine ecosystem, and then the interest in minerals and the minerals regime now under negotiation. Two critical issues are highlighted: first, whether the agreements being negotiated by the small group of Treaty powers will deal wisely with resource issues, and spur the needed scientific research on resources questions; second, the issue of international equity—that is, whether developing nations, which are becoming more vocal about their wish to have a say in Antarctica's disposition; can indeed have a role in the region's governance. The final chapter discusses possible ways that the region's political system could evolve in the future, before and after 1991 when the Treaty may be reviewed.

Fernando Zegers Santa Crus (1978). The Antarctic system and the utilization of resources. *University of Miami Law Review*, **33**, 47 pp. Original Spanish version printed alongside.

The author discusses the legal status of Antarctica in the context of resources utilization. He maintains that Antarctic Treaty provisions and the resultant activities by Treaty Parties have been at least tacitly accepted by the international community. Although the Treaty has not been signed by the majority of the world's nations, most of those which have expressed an interest in the area are signatories or have acceded to it. He describes the negotiation of the Antarctic Treaty, the operation of the Treaty system, environmental protection and conservation of natural resouces under it, the question of sovereignty and the problems of utilization of resources. In relation to sovereignty he affirms the Chilean view and notes that any proposal for the utilization of Antarctic resources will have to take claims into account. He concludes that the solution to resources utilization in Antarctica must be found through Antarctic Treaty system and in close cooperation with it.

REFERENCES

Adam, K. M. (1979). Building and operating winter roads in Canada and Alaska. Environmental Studies no. 4. Dept. of Indian and Northern Affairs Report R71-19/4-1978, 122 pp.

AEIMEE (1981, 1983). Reports nos. 1 and 2 of SCAR committee of specialists on Antarctic environmental implications of possible mineral exploration and exploitation. Unpublished reports, SCAR secretarial headquarters, Scott Polar Research Institute, Cambridge, UK.

Agterberg, F. P. (1981). Application of image analysis and multivariate analysis to mineral resource appraisal. *Econ. Geol.* **76**, 1016–31.

Agterberg, F. P. and Divi, S. R. (1978). A statistical model for the distribution of copper, lead and zinc in the Canadian Appalachian region. *Econ. Geol.* **73**, 230–45.

Agterberg, F. P., Chung, C. F., Divi, S. R., Rade, K. E., and Fabbri, A. G. (1981). Preliminary geomathematical analysis of geological, mineral occurrence and geophysical data, southern district of Keewatin, North West Territories, Geological Survey of Canada, Open file 718.

Alexander, A. (1978). Recommended approach to the Antarctic Resource problem. *U. Miami Law Rev.* **33**, 417–23.

Ali, S. and Richardson, M. (1983). An eye on Antarctica. India is continuing its scientific expenditions to the southern continent, in part to ensure a future stake there. *Far East Econ. Rev.*, pp. 32–3.

Allais, M. (1957). Method of appraising economic prospect of mining exploration over large territories. Algerian Sahara case study. *Manage. Sci.* **3**, 285–347.

Amundsen, R. (1912). *The South Pole: an account of the Norwegian expedition in the 'Framn'* (1910–1912). John Murray, London, 2 vols. Reproduced in 1976 by Harper and Row, Barnes and Noble Division, New York.

Anonymous (1956). Minerals of the Antarctic. *Min. J.* **246**, 149–50.

Anonymous (1978). Thaw in international law? Rights in Antarctica under the law of common spaces. *Yale Law J.* **87**, 804–59.

Anderson, J. B., Kurt, D., Weaver, F., and Weaver, M. (1982). Sedimentation on the West Antarctic continental margin. In *Antarctic Geosciences, Geology and Geophysics* (ed. Campbell Cradock) pp. 1103–12. IUGS series B, No. 4, University of Wisconsin Press, Madison WI.

Andor, L. E. (1985). *South Africa's Chrome, Manganese, Platinum and Vanadium. Foreign views on the mineral dependency issue,* 1970–1984. A select and annotated bibliography. South African Inst. of International Affairs, Bibliographical Series, No. 13, 222 pp.

Antarctic (1975). Filchner ice shelf station's major project this summer. *Antarctic* **7**, 211.

Antarctic (1978). 265 men wintering at six permanent stations. *Antarctic* **8**, 189.

Antarctic (1979*a*). Research in the Dufek Massif area likely. *Antarctic* **8**, 187.

Antarctic (1979*b*). West German design for first base. *Antarctic* **8**, 350.

Antarctic (1980). Soviet plan for another winter station. *Antarctic* **9**, 56.

Antarctic (1981). Reference to Japanese trials at Syowa, with the Misui hovercraft, for Antarctic logistics. *Antarctic* **9**, 276.

Antarctic (1982). Reference to Argentine mineral exploration on Antarctic Peninsula. *Antarctic* **9**, 359.

Antarctica—Memorandum (1983). Memorandum by the delegation of Malaysia on Antarctica as a common heritage of mankind, 7th Conference of Heads of State of Governments of Non-alligned Countries, New Delhi, 3 March (NAC/CONF.7/INF.11).

Antarctic Journal of the United States (1980). McMurdo Station reactor site released for unrestricted use. *Antarctic J. U.S.* **15**, 1–4.

Antarctic Journal of the United States (1982). Uranium resources evaluation; resource and radioactivity survey by airborne gamma ray spectrometry. *Antarctic J. U.S.* **17**, 9–10.

Antarctic Journal of the United States (1983). Williams Field; the history of an icy aerodrome; Williams Field today. *Antarctic J. U.S.* **18**, 20–1.

Anti-Apartheid Movement (1981). *Apartheid gold.* Anti-Apartheid Movement and UN Centre against Apartheid,, London, 21 pp.

Archer, A. A. (1983). Marine mineral resources. Effects of the law of the Sea Convention. *Resources Policy* March, 23–32.

Argentina National Report to SCAR (1981). Scientific activities for years 1980–1981 and planned program for 1981–1982. Progress report No. 23. Instituto Antarctico Argentino.

Arrow, K. J. and Chang (1978). Optimal pricing, use and exploitation of uncertain natural resource stocks. Technical report No. 31, Dept. of Economics, Harvard University.

ASOC (1984). The Antarctic and Southern Ocean Coalition. Report on the Antarctic mineral meeting, Washington DC, USA, 18–27 January 1984.

Atkins, F. B. (1969). Pyroxenes of the Bushveld Intrusion, South Africa. *J. Petrol.* **10**, 222–49.

Auburn, F. M. (1982). *Antarctic law and politics.* Hurst & Co., London, 361 pp.

Auburn, F. M. (1984). Antarctic minerals and the third world. *J. Polar Stud.* **1**, 201–223.

Aughenbaugh, N. B. (1961). Preliminary report on the geology of the Dufek Massif: International Geophysical Year World Data Centre A, Glaciology Report No. 4, pp. 155–93.

Australian Department of Foreign Affairs (1983). Handbook of Measures in furtherance of the principles and objectives of the Antarctic Treaty. Canberra (1983).

Australian Department of Science and Technology (1982). Antarctic Division Annual Review, (1981–1982). Australian Government Publishing Service, Canberra, 28 pp.

Barnet, R. J. (1980). *The lean years. Politics in the age of scarcity.* A Touchstone book, published by Simon and Schuster, New York, 349 pp.

Beck, M. E. (1972). Paleomagnetism and magnetic polarity zones in the Jurassic Dufek Intrusion, Pensacola Mountains, Antarctica. *Geophys. J.R. Astron. Soc.* **28**, 49–63.

Beck, M. E., Burmester, R. F., and Sheriff, S. D. (1979). Field reversals and palaeomagnetic pole for Jurassic Antarctica, EOS, *Trans. Am. Geophys. Union* **60**, 818.

Beck, P. J. (1984). Antarctica: a case for the U.N.? In *The world today*, pp. 165–71 Chatham House, London. See also this author's forthcoming book *The international politics of Antarctica.*

Beeby, C. D. (1983). An overview of the problems which should be addressed in the preparations of a regime governing the mineral resources of Antarctica. In *Antarctic resources policy, scientific, legal and political issues.* (ed. F. O. Vicuña) pp. 191–8. Cambridge University Press, Cambridge, UK.

Beenstock, M. (1983). *The world economy in transition.* George Allen and Unwin, London, 238 pp.

Behrendt, J. G. (1979). Speculations on petroleum resource potential of Antarctica. *Am. Ass. Petr. Geol. Bull.* **63**, 418.

Behrendt, J. G., ed. (1983). Petroleum and mineral resources of Antarctica. US Geological Survey Circular No. 909, 75 pp.

Behrendt, J. C., Drewry, D. J., Jankowski, E., and Grim, M. S. (1980). Aeromagnetic and radio echo ice-sounding measurements show much greater area of the Dufek Intrustion Antarctica. *Science* **209**, 1014–16.

Behrendt, J. G., Henderson, J. R., Meister, L., and Rambo, W. L. (1974). Geophysical investigations of the Pensacola Mountains and adjacent glacierized areas of Antarctica. U.S. Geological Professional paper No. 844, 28 pp.

Boletin Antarctico Chileno (1982). Cuadro resumen Programa Cientifico Expedition No. 19 (1982/1983). *Bol. Antarctico Chil.* **2**, 3.

Bosson, R. and Varon, B. (1977). *The mining industry and the developing countries.* Oxford University Press, Oxford. 292 pp.

Boswell, P. G. (1982). The high temperature stress–rupture properties of platinum and palladium. The effect of environment and composition on service performance. *Platinum Met. Rev.* **26**, 16–19.

Boulding, K. E. (1983). Ecodynamics. *Interdiscip. Sci. Rev.* **8**, 108–113.

Bormann, P. and Weber, W. (1983). Mineral resources of Antarctica in the light of Gondwana geology. IUGG Inter-Disciplinary Symposia Hamburg, Programme and Abstracts, Vol. 1, p. 222.

Blanchard, F. (Director General, ILO) (1976). Employment, growth and basic needs. A one-world problem. Tripartite World Conference on employment, income distribution and social progress and the international division of labour. Report of the Director-General of the ILO and declarations of principles and programme of action adopted at the conference. International Labour Office, Geneva, 211 pp.

Brandt, W. (1980). *North–south: a programme for survival.* Report of the Independent Commission on International Development Issues. MIT Press, Cambridge, MA, 304 pp.

Brandt, W. (1983). *Common crisis, North–south: co-operation for world recovery.* Pan Books, London, 174 pp.

Brennan, K. (1983*a*). International constraints. In: The Antarctic: preferred futures, constraints and choices. Proceedings of a seminar held by the New Zealand Institute of International Affairs, New Zealand, Wellington, June 1983. N.Z. Inst. of International Affairs, Wellington, pamphlet No. 44, pp. 12–24.

Brennan, K. (1983*b*). Criteria for access to the resources of Antarctica: alternatives, procedure and experience applicable. In: *Antarctic resources policy. Scientific, legal and political issues* (ed. F. O. Vicuña) pp. 217–227. Cambridge University Press, Cambridge, UK.

Brewster, B. (1982). *Antarctica, wilderness at risk.* Friends of the Earth Books, San Francisco, 125 pp.

British Antarctic Survey (1982). Annual Report 1981–1982. Natural Environment Research Council. BAS, Cambridge, UK, 83 pp.

British Antarctic Survey (1979–1983). Geological Map Series 1: 500,000 (BAS 500G), sheets 1–5.

Broecker, W. S. (1984). Carbon dioxide circulation through ocean and atmosphere. *Nature* **308**, 602.

Brown, E. D. (1983). Deep sea mining: The consequences of failure to agree at UNCLOS III. Natural Resources Forum, United Nations, No. 7, pp. 55–70.

Buchanan, D. L. (1978). The economic potential and petrography of Bushveld rocks in the Bethal area. Bureau for Mineral Studies, Johannesburg, Report No. 2, 10 pp.

Buchanan, D. L. (1979*a*). Platinum—great importance of Bushveld complex. *World Min.* August, pp. 56–9.

Buchanan, D. L. (1979*b*). Platinum—group metal production from the Bushveld complex and its relationship to world markets. Bureau for Mineral Studies, Johannesburg, Report No. 4, 31 pp.

Buchanan, D. L. (1981). Strategic minerals and the European Community. Background paper to proposed primary raw materials enquiry, House of Lords, 8 pp.

Buchanan, D. L. (1982). Western resources of platinum-group metals, nickel and chromium: their production and future demand. In: *Proceedings of the 12th CMMI Congress* (ed. H. W. Glen) pp. 259–62. South African Institute of Mining and Metallurgy, Johannesburg.

Buchanan, D. L. and Nolan, J. (1979). Solubility of sulphur and sulphide immiscibility in synthetic tholeiitic melts and their relevance to Bushveld complex rocks. *Can. Mineral.* **17**, 483–94.

Buchanan, D. L. and Rouse, J. E. (1984). Role of contamination in the precipitation of sulphides in the Platreef of the Bushveld Complex. In *Sulphide deposits in mafic and ultramafic rocks* (ed. D. L. Buchanan and M. J. Jones) pp. 141–6. Special Volume of the Institute of Mining and Metallurgy, London.

Burmester, R. F. and Sheriff, S. D. (1980). Palaeomagnetism of the Dufek Intrusion, Pensacola Mountains, Antarctica. *Antarctic J. U.S.* **15**, 43–5.

Bush, W. M. (1982). *Antarctica and international law. A collection of interstate and national documents.* Vol. 1. Oceans Publications, London, 703 pp.

Cabri, L. J. (1981*a*). Nature and distribution of platinum-group element deposits. *Episodes* **2**, 31–34.

Cabri, L. J. (1981*b*). *Platinum-group elements: mineralogy, geology, recovery,* p. 267. Canadian Institute of Mining Special Volume 23.

Calder, N. (1983). *1984 and after.* Changing images of the future. Century Publishing, London, 207 pp.

Cameron, D. S. (1981). Progress in fuel cell technology. A report of the United States National Seminar. *Platinum Met. Rev.* **25**, 161–2.

Campbell, I. H. (1984). Layered complexes: mineralisation and mechanisms. A two-day course, University of the Witwatersrand, unpublished course notes, 69 pp.

Campbell, I. H., Naldrett, A. J., and Barnes, S. J. (1983). A model for the origin of the platinum-rich sulfide horizons in the Bushveld and Stillwater complexes. *J. Petrol.* **24**, 133–65.

Cannon, W. T. (1979). New mining ventures: the case for preserving after-tax financing. Centre for Resource Studies, Queen's University, Kingston, Canada, 19 pp.

Cawthorn, R. G. and McCarthy, T. S. (1980). Variations in Cr content of magnetite from the upper zone of the Bushveld complex—evidence for heterogeneity and convection currents in magma chambers. *Earth Planet. Sci. Lett.* **46**, 335–43.

Chalmers, R. O. (1957). Mineral possibilities of the Antarctic. *Aust. Mus. Mag.* **2** (8), 252–55.

Chamber of Mines of South Africa (1980). South Africa as a supplier of minerals to the West. Chamber of Mines of South Africa Publications No. 3, 6 pp.

Chaston, J. C. (1982). The growing industrial use of the platinum metals. A quarter century of technological progress. *Platinum Met. Rev.* **26**, 3–90.

CIA (1978). Polar regions atlas. Central Intelligence Agency, Washington. Updated 1981, 1983.

Christian, J. M. (1983). Platinum and palladium. A third year of depressed demand keeps markets weak. *Eng. Min. J.* March, 65–7.

Codling, R. J. (1982). Sea-borne tourism in the Antarctic—an evaluation. *Polar Rec.* **21**, 3–9.

Cohen, R. S. and O'Nions, R. K. (1982*a*). The lead, neodymium and strontium isotopic structure of Ocean Ridge basalts. *J. Petrol.* **23**, 299–324.

Cohen, R. S. and O'Nions, R. K. (1982*b*). Identification of recycled continental material in the mantle from Sr, Nd, and Pb isotope investigations. *Earth Planet. Sci. Lett.* **61**, 73–84.

Cominco Ltd (1979, 1980, 1981). Annual Reports.

Cominco Ltd Northern Group (1979). Polaris mine project summary report, July, unpublished.

Conn, H. K. (1979). The Johns–Manville Platinum–Palladium Prospect, Stillwater Complex, Montana, USA. *Can. Mineralogist* **17**, 463–8.

Cook, Captain, J. (1777). A voyage to the South Pole and round the world (2 vols.) London.

Cook, F. A. (1983). Modelling the Ross Sea basins for hydrocarbon evaluation. N.Z. Antarctic Records No. 5, Pacific Science Conference, 15th High Latitude Symposium, p. 16–24.

Corrans, I. J., Brugman, C. F., Overbeek, P. W., and McRae, L. B. (1982). The recovery of platinum-group metals from ore of the UG-2 reef in the Bushveld Complex. *Proceedings of the 12th CMMI Congress* (ed. H. W. Glen) pp. 629–34. South African Institute of Mining and Metallurgy, Johannesburg.

Coupland, D. R., Hall, C. W., and McGill, I. R. (1982). Platinum-enriched superalloys. A development alloy for use in industrial and marine gas turbine environments. *Platinum Met. Rev.* **26**, 146–57.

Couratier, J. (1983). The regime for the conservation of Antarctica's living resources. In: *Antarctic resources policies. Scientific, legal and political issues* (ed. F. O. Vicuña) pp. 139–48. Cambridge University Press, Cambridge, UK.

Current Antarctic literature and Antarctic bibliography (1951–1983). Prepared by the cold regions bibliography project science and technology division for the division of polar programs, NSF. Library of Congress, Washington DC.

Dalziel, I. W. D. (1983). Geologic transect across the southernmost Chilean Andes: report of R/V *Hero* Cruise 84-4. *Antarctic J. U.S.* **18**, 8–12.

Dasgupta, P. S. and Heal, G. M. (1979). *Economic theory and exhaustible resources.* Cambridge University Press, Cambridge UK, 501 pp.

Davies, G. and Tredoux (1985). The platinum group element and gold content of the marginal rocks and sills of the Bushveld Complex. *Econ. Geol.* **80,** (in press).

De Paolo, D. J. (1983). The mean life of continents: Estimates of continental recycling rates from Nd and Hf isotopic data and implications for mantle structure. *Geophys. Res. Letters* **10,** 705–8.

De Waal, S. A. (1977). Carbon dioxide and water from metamorphic reactions as agents for sulphide and spinnel precipitation in mafic magmas. *Trans. Geol. Soc. S. Afr.* **80**, 193–6.

de Wit, M. J. (1981). Assessment of Antarctica's mineral resource potential and the related economic projections of costs, risks and returns associated with their discovery, development and exploitation. Unpublished research proposal. Dept. of Geological Sciences and Centre of Resources Studies, Queen's University, Kingston, Canada.

de Wit, M. J. and Bergh, H. (1984). Antarctic mineral resources—studied in the framework of geological and tectonic–metal anomaly maps of Gondwanaland—in progress.

de Wit, M. J., Jeffery, M., Bergh, H., and Nicolaysen, L. O. (1985). *Geologic map of sectors of Gondwana reconstructed to their disposition* c. *180 m.y.* American Association of Petroleum Geologists, Tulsa, OK (in press).

de Wit, M. J., Spencer, R., and Buchanan, D. (1984). Technical and economic feasibilities of platinum mining in Antarctica. Occasional paper, Institute of Mining and Metallurgy, London (in preparation).

De Young, Jr. J. H. (1981). The Lasky cumulative tonnage–grade relationship—a re-examination. *Econ. Geol.* **76**, 1067–80.

Dixon, C. J. (1979). *Atlas of economic mineral deposits.* Chapman and Hall, London, 143 pp.

Diwell, A. F., and Harrison, B. (1981). Car exhaust catalyst for Europe. The development of lead tolerant platinum catalyst systems for emission control. *Platinum Met. Rev.* **25**, 142–51.

Drewry, D. J. (1982). Antarctica unveiled. *New Sci.* 22 July, 246–51.

Drewry, D. J., ed. (1983). Antarctica glaciological and geophysical folio. Scott Polar Research Institute, University of Cambridge, UK.

Doyle, E. N. (1980). Environmental controls: In: Northern Mining in the 1980s. Centre for Resource studies, Queen's University, Kingston, Ontario. Proceedings No. 10 of the NWT Chamber of Mines. Mining Days 1980. Yellowknife May 1980, pp. 41–6.

Dreschhoff, G. A. M., Zeller, E. J., Schmid, H., Bulta, K., Morency, M., and Tremblay, A. (1983). Radioactive mineral occurrence at Szabo Bluff, Transantarctic mountains. *Antarctic J. US* **18**, 48–9.

Dugger, J. A. (1974). Exploiting Antarctic mineral reserves. Technology, economics of the environment. *Univ. Miami Law Rev.* **33**, 315–39.

Du Toit, A. L. (1937). *Our wandering continents.* Oliver and Boyd, Edinburgh, 366 pp.

EAMREA (1979). *Possible environmental effects of mineral exploration and exploitation in Antarctica* (ed. J. H. Zumberge). SCAR publication, Cambridge, 59 pp.

Economic Commission on Europe (1983). Public hearings on acid deposition (acid rain) by the European Parliament Committee on the Environment, public health, consumer protection, Brussels. (April 1982).

ECO (1983). Antarctic minerals regime—Beeby's slick solution. Special publication to cover the Special Consultative meeting on Antarctic Mineral Resources, Bonn, July 11–22, 1983. Friends of the Earth International, Washington. *ECO,* **13,** 15 pp.

ECO (1984*a*). Antarctic minerals negotiations continue. Special publication to cover the Special Consultative Meeting on Antarctic Mineral Resources, Washington, DC, January 18–27, 1984. *ECO,* **26,** Nos. 1–3, 24 pp.

ECO (1984*b*). Antarctic Minerals Negotiations continue. Special publication to cover the special consultative meeting on Antarctic Mineral Resources, Tokyo, 22–31 May 1984. *ECO* **27,** Nos 1–3, 20 pp.

Economic Geology Subdivision, Geological Survey, Canada (1978). Evaluation of the regional mineral potential (non-hydrocarbon) of the Western Arctic region. Open file No. 492, Ottawa, 31 pp.

Economic Geology Division (1980). Non-hydrocarbon mineral resource potential of parts of Northern Canada. Preliminary resource assessments of parts of Northern Yukon, Mainland Northwest Territories and the Arctic Islands, including islands in Hudson Bay. Geological Survey of Canada, open file No. 716. part I Methodology and summary assessments, Part II Detailed assessments and appendixes; 376 pp.

Elliot, D. H. (1976). A framework for assessing environmental impacts of possible Antarctic mineral development. (Including the economics of hypothetical mining ventures.) Ohio State University, Institute of Polar Studies. Unpublished report, 620 pp.

Elliot, D. H. (1983). Earth science investigations in the United States Antarctic research program for the period July 1, 1982–June 30, 1983. National Academy Press, Washington, DC, 91 pp.

Elliot, D. H. (1984). Southern Ocean Drilling. SCAR bulletin No. 77, pp. 30–32.

Emmel, P. G. (1984). Platinum group catalysts for Europe. Car emission control and future metal demand. *Platinum Met. Rev.* **28**, 22–24.

Etheridge, D. (1980). Address to London Metal Forum, quoted in *Rand Daily Mail,* 14 October, p. 15.

Ericksen, G. E. (1976). Metallogenic provinces of south eastern pacific region. In: *Circum-Pacific energy and mineral resources* (ed. M. T. Halbouty, J. C. Maher and H. M. Lian, pp. 527–38. American Association of Petroleum Geologists, Memoir No. 25.

Europa Year Book (1983). A world survey, Vol. 1. Europe Publications Ltd, London, 1790 pp.

Fairbridge, R. W. (1952). The geology of the Antarctic. In: *The Antarctic Today* (ed. F. A.

Simpson, A. H. Wellington, and A. W. Reed) pp. 56–101. New Zealand Antarctic Society.

Fastook, J. L. and Hughes, T. (1982). When ice sheets collapse. Perspectives in computing. *Appl. Acad. Sci. Community* **2**, 4–15.

Federal Republic of Germany (1983). National Antarctic Research Report (4) to SCAR. Activities 1981–1982; planned activities 1982–1983.

Fenge, T., Gardner, J., E., King, J., and Wilson, B. (1979). Environment Canada. Land Use programmes in Canada. NWT. Faculty of Environmental Studies, University of Waterloo. Minister of Supply and Services, Ottawa, 296 pp.

Fermor, L. L. (1951). The mineral deposits of Gondwanaland. *Trans. Inst. Min. Metall. Lond.* **60**, 421–65.

Fifield, R. (1982). Antarctic scientists buoyed up by South Atlantic cash. *New Sci.* 18 Nov., **96**, p. 406.

Fish, R. (1974). The Black Angel experience. *Can. Min. J.* 26–36.

Ford, A. B. (1970). Development of the layered series and capping granophyre of the Dufek intrusion of Antarctica. *Geol. Soc. S. Afr.,* Special Publication No. 11, pp. 492–510.

Ford, A. B. (1976). Stratigraphy of the layered gabbroic Dufek intrusion. Antarctica. *U.S. Geol. Surv. Bull.* **1405-D**, 36 pp.

Ford, A. B. (1983). The Dufek intrusion of Antarctica and a survey of its minor metals and possible resources. In (ed. Behrendt, J. A.) *Petroleum and mineral resources of Antarctica,* pp. 51–75. Geological Survey of US, Circular No. 909.

Ford and Boyd (1968). The Dufek intrusion, a major stratiform gabbroic body in the Pensacola Mountains, Antarctica. XXIII International Geological Congress, Prague, Vol. 2, pp. 213–28.

Ford, A. B. and Kistler (1980). K–Ar age, composition and origin of Mesozoic mafic rocks related to Ferrar Group, Pensecola Mountains, Antarctica. *N. Z. Geol. Geophys.* **23**, 371–90.

Ford, A. B., Schmidt, D. L., and Boyd, W. W. (1978*a*). Geology of the Saratoga table quadrangle Pensacola Mountains, Antarctica. Notes to accompany map A-9. Department of the Interior, United States Geological Survey, 9 pp.

Ford, A. B., Schmidt, D. L., and Boyd, W. W. (1978*b*). Geology of the Davis valley quadrangle and parts of the Cordiner peaks quadrangle, Pensacola mountains, Antarctica. Notes to accompany Map A-10, 10 pp.

Ford, A. B., Reynolds, R. L., and Huie, C. (1979). Geological field investigations of Dufek intrusion. *Antarctic J. US* **14**, 9–11.

Ford, A. B., Carlson, C., Czamanske, G. K., Nelson, W. H., and Nutt, C. J. (1977). Geological studies of the Dufek intrusion, Pensacola Mountains (1976–1977). *Antarctic J. US* **12**, 90–92.

Fuchs, Sir V. and Hillary, Sir E. (1958). *The crossing of Antarctica.* Cassell, London, 338 pp.

Fuchs, V. E. (1983). Trans-Antarctic Expedition (1955–1958. Scientific reports No. 1: synopsis of results. Trans-Antarctic Association publication, Cambridge, UK, 14 pp.

Geldenhuys, D. (1981). Some foreign policy implications of South Africa's 'Total National Strategy' with particular reference to the '12-point plan'. South African Institute of International Affairs, 63 pp.

German Democratic Republic: Antarctic research activities (1982)., Report to SCAR 3. Activities 1981–1982; planned activities 1982–1983.

German Tribune (1983). Antarctic eco-system endangered as scientists move in. January, No. 1070, pp. 9–10. Originally published by Gert Kistenmacher, *Suddeutsche Zeitung* (14 January 1983).

Gjelsvik, T. (1983). The mineral resources of Antarctica: progress in their identification. In: *Antarctic Resources Policy: Scientific, legal and political issues* (ed. F. O. Vicuña), pp. 61–76. Cambridge University Press, Cambridge, UK.

Glacken, I. M. (1981). The platinum group metals; geology and profitability. Unpublished thesis, Imperial College, University of London, 127 pp.

Goldstein, W. (1982). Western concerns over seabed mining. *Resources Policy* June, 82–83.

Goldstein, W. (1983). No treaty code for the seabed. *Resources Policy* March, 2–3.

Gordon, A. L. (1981). Seasonality of Southern Ocean sea ice. *J. Geophys. Res.* **86**, 4193–97.

Graham, K. A. (1981). Eastern Arctic case study series: the development of the Polaris Mine. Centre for Resource Studies, Queens University Kingston, Ontario, Canada. Preliminary draft, 69 pp.

Graham, K. A., McEachern, R. G., and Miller, C.

G. (1979). The administration of mineral exploration in the Yukon and Northwest Territories. Centre for Resource Studies, Queen's University, Kingston, Canada. Working Paper No. 14, 42 pp.

Greenpeace (1984). An Antarctic environmental protection agency. Antarctic briefing No. 6, Greenpeace International Council and the Antarctic Project, Washington DC. 2 pp.

Gregory, M. R. (1982). Ross Sea hydrocarbon prospects and the IXTOC I oil blow-out Campeche Bay, Gulf of Mexico, comparisons and lessons. *N.Z. Antarctic Rec.* **4**, 40–5.

Grew, E. S. and Manton, W. I. (1979). Archaean Rocks in Antarctica: 2.5 billion-year uranium–lead ages of pegmatites in Enderby Land. *Science* **206**, 443–5.

Groves, D. I., Lesher, C. M., and Gee, R. P. (1984). Tectonic setting of the sulphide nickel deposits of the Western Australian Shield. In: *Sulphide deposits in mafic and ultramafic rocks* (ed. D. Buchanan and M. Jones) pp. 1–13. Institute of Mining and Metallurgy, London.

Hansard (1983). House of Lords official report parliamentary debates. EEC 20th Report: strategic minerals, 21 February, Vol. 439, pp. 548–76.

Harris, D. P. (1984). Mineral resources appraisal. *Mineral endowment, resources, and potential supply: concepts, methods, and cases.* Oxford Geological Science Series I. Clarendon Press, Oxford. 445 pp.

Harris, D. P., Freyman, A. J., and Barry, G. S. (1970). The methodology employed to estimate potential mineral supply of the Canadian Northwest—an analysis based upon geologic opinion and systems simulation. *Can. Mineral Inf. Bull.* **MR105**, 1–56.

Hartwick, J. M. (1983). Learning about and exploiting exhaustible resource deposits of uncertain size. *Can. J. Econ.* **16**, 391–410.

Hattingh, P. J. (1983). A palaeomagnetic investigation of the layered mafic sequence of the Bushveld complex. University of Pretoria, Ph.D. thesis, unpublished, 177 pp.

Haughton, D. R., Roeder, P. L., and Skinner, B. J. (1974). Solubility of sulphur in mafic magmas. *Econ. Geol.* **69**, 451–67.

Hawkes, D. D. and Littlefair, M. J. (1982). An occurrence of molybdenum, copper and iron mineralisation in the Argentine Islands, West Antarctica. *Econ. Geol.* **76**, 898–904.

Hayfield, P. C. S. (1983). Platinised titanium electrodes for cathodic protection. *Platinum Met. Rev.* **27**, 2–8.

Heap, J. (1983). Question of Antarctica. 38th Session Committee, Agenda Item 140, United Nations General Assembly A/C.1/38/L.80, 28 November 1983. Limited Distribution, pp. 21–6.

Herbert, M. T., Acres, G. J. K., and Hughes, J. E. (1980). Platinum, gold, silver and mercury. In: *Future metal strategy. Proceedings of an international conference organised by the Metals Society.* May 1979, pp. 109–24. The Metals Society, London.

Hewish, M. (1983). Quiet craft hovers ahead of competition. *New Sci.* **97**, 297–99.

Himmelberg, G. R. and Ford, A. B. (1976). Pyroxenes of the Dufek intrusion, Antarctica. *J. Petrol.* **17**, 219–43.

Himmelberg, G. R. and Ford, A. B. (1977). Iron–titanium oxides of the Dufek intrusion, Antarctica. *Am. Mineral.* **62**, 623–33.

Himmelberg, G. R. and Ford, A. B. (1983). Composite inclusion of olivine gabbro and calc-silicate rock in the Dufek intrusion, a possible fragment of a concealed contact zone. *Antarctic J. US*, **18**, 1–4.

Hinz, K. and Krause, W. (1982). The continental margin of Queen Maud Land/Antarctica: seismic sequences, structural elements and geological developments. *Geol. Jhb.* **23**, 17–41.

Hofmann, A. W. (1984). Geochemical mantle models. *Terra cognita* **4**, 157–66.

Holdgate, M. W. and Tinker, J. (1979). Oil and other minerals in the Antarctic. The environmental implications of possible mineral exploration or exploitation in Antarctica. London, House of Print, 51 pp.

Holdgate, M. W. (1983*a*). Policy for Antarctica resources. *Polar Rec.* **21**, 342–93.

Holdgate, M. W. (1983*b*). Environmental factors in the development of Antarctica. In: *Antarctic resources policy, scientific, legal and political issues* (ed. F. O. Vicuña), pp. 77–101. Cambridge University Press.

Hotelling, H. (1931). The economics of exhaustible resources. *J. Polit. Econ.* **39**, 137–75.

House of Lords (1982). Strategic minerals. Select committee on the European communities, session (1981–1982), 20th report, 328 pp.

Huffman, J. W. and Schmidt, D. L. (1966). Pensacola mountain project. *Antarctic J. US* **1**, 123–4.

Indian and Northern Affairs, Canada (1981). Mines and minerals activities (1980). Canadian Government, Department of Indian and Northern Affairs, Ottawa, 33 pp.

Infanta Caffi, M. T. (1982). Seminario sobre politica para los recursos antarticos. *Bol. Antartico Chil.* **2**, 42–5.

Innes, D. (1984). *Anglo-American and the rise of modern South Africa.* Ravan Press, Johannesburg, 358 pp.

IUCN (1972). Second World Conference on National Parks meeting at the Grand Teton National Park, USA (September 1972). Quoted in Mitchell (1982), pp. 137 and 180.

Ivanhoe, L. F. (1980). Antarctica: Operating conditions and petroleum prospects. *Oil Gas J.* **78**, 212–20.

Janis, I. L. (1972: revised 1982). *Victims of Groupthink: a psychological study of foreign-policy decisions and fiascoes.* Houghton Mifflin, Boston, 277 pp.

Janis, I. L. and Mann, L. (1977). *Decision-making. a psychological analysis of conflict, choice and commitment.* Free Press, New York, 488 pp.

Jenks, C. W. (1958). *The common law of mankind,* pp. 365–81. Prager, New York.

Johnson Matthey (1982*a*). Johnson Matthey technology world wide. Platinum metals fulfil evolving industrial needs. *Platinum Met. Rev.* **26**, 105.

Johnson Matthey (1982*b*). Catalysts for Europe. Lead-tolerant catalysts. Unpublished report, JMC Ltd., Royston, UK, 16 pp.

Johnson Matthey (1983*a*). Catalysts for Europe: availability. Unpublished report, JMC Ltd, Royston, UK, 5 pp.

Johnson Matthey (1983*b*). Reports on public hearings on acid deposition (acid rain) by the European Parliament Committee on the Environment, public health and consumer protection, Brussels April 1983. Unpublished Reports, JMC Ltd, Royston, UK.

Jolly, J. M. (1978). Platinum group metals. US Bureau of Mines Mineral Commodity Profiles, No. MCP-22.

Kameneva, G. I. and Grikurov, G. E. (1983). A metallogenic reconnaissance of Antarctic major structural provinces. In: *Antarctic earth science* (ed. R. L. Oliver, P. R. James, and J. B. Jago), pp. 42–43. Australian Academy of Science, Canberra.

Kanehira, K. (1979). A note on assessment of mineral resources in the Antarctic. In: Proceedings of the First Symposium on Antarctic Geosciences (1978). Memoirs of National Institute of Polar Research, Special Issue No. 14.

Karig, D. E. and Kay, R. W. (1981). Fate of sediments on the descending plate at convergent margins. *Phil. Trans. R. Soc. Lond.* **A301**, 233–51.

Kay, R. (1980). Volcanic arc magmas: implications of a melting–mixing model for element recycling in the crust-upper mantle system. *J. Geol.* **88**, 497–522.

Kay, R. (1985). Island arc processes relevant to crustal and mantle evolution. *Tectonophysics* (in press).

Kimball, L. (1983*a*). Report on Antarctic events (1983). International Institute for Environment and Development (IIED), Washington, DC. 19 pp.

Kimball, L. (1983*b*). Antarctica: a continent in transition. A portfolio. IIED, publication, London, UK.

Knox, G. A. (1983). The living resources of the Southern Ocean: a scientific overview. In: *Antarctic resources policy: scientific, legal and political issues* (ed. F. O. Vicuña), pp. 22–59. Cambridge University Press, Cambridge, UK.

Krapels, E. N. (1980). *Oil crisis management, strategic stockpiling for international security.* Johns Hopkins University Press, Baltimore.

Kronmiller, T. G. (1980). *The lawfulness of deep seabed mining.* Oceana Publications, London, 2 Vols, 980 pp.

Kruger, F. J. and Marsh, J. S. (1982). The significance of $^{87}Sr/^{86}Sr$ ratios in the Merensky Cyclic unit of the Bushveld complex. *Nature* **298**, 53–5.

Kruger, F. J. and Marsh, J. S. (1985). The mineralogy, petrology and origin of the Merensky cyclic unit in the western Bushveld complex. *Econ. Geol.* **80**(in press).

La Grange, J. J. (1963). Meteorology: (1) Shackleton, Southice and the journey across Antarctica. Trans-Antarctic Expedition Scientific Reports No. 13. Transactions of the Antarctic Association, Cambridge, UK.

Leggatt, C. H. (1982). Polaris—world's most northerly mine. *World Min.* September, 46–51.

Lewis, T. R. and Schmelemsee, R. (1980). On oligopolistic markets for non-renewable natural resources. *Q. J. Econ.* **95**, 475–91.

Lipps, J. H. (1978). Man's impact along the Antarctic peninsula. In: *Environmental impact in Antarctica* (ed. B. C. Parker), pp. 333–71. Virginia Polytechnic Institute and State University, Blacksburg, Virginia.

Lister, H. (1960). Solid precipitation and drift snow. Trans-Antarctic expedition scientific report No 5: Glaciology. Trans-Antarctic Association, Cambridge UK.

Lock, R. G. (1983). Continental margin petroleum potential in the Ross Sea region. *N.Z. Antarctic Rec.* **5**. Pacific Science Conference, 15th High Latitude Symposium, pp. 6–15.

Lovering, J. F. and Prescott, J. R. V. (1979). *Last of Lands, Antarctica.* Melbourne University Press, 212 pp.

Lyons, L. A. (1983). Discovery and development of Australian diamonds. *World Min.* September, 50–56.

Mackenzie, B. W. (1980). Looking for the improbable needle in a haystack: the economics of base metal exploration in Canada. Centre for Resource Studies, Queens University, Kingston, Working paper no. 19, 43 pp.

Mackenzie, B. W. and Bilodean, M. L. (1979). Effects of taxation on base metal mining in Canada. Centre for Resource Studies, Queen's University, Kingston, Ontario, Canada, 190 pp.

Marston, R. D., Groves, D. I., Hudson, D. I., and Ross, J. R. (1981). Nickel sulfide deposits in W. Australia: a review. *Econ. Geol.* **76**, 1330–63.

Matsueda, H., Motoyoshi, Y., and Matsumoto, Y. (1983). Mg–Al skarn of the Skallevikhalsen on the east coast of Lutzow-holn Bay, East Antarctica. Proceedings of the Third Symposium on Antarctic Geosciences, 1982, pp. 166–182.

McIntosh, W. C., Kyle, P. R., Cherry, E. M., and Noltimer, H. C. (1982). Palaeomagnetic results from the Kirkpatrick basalt group, Victoria Land. *Antarctic J. U.S.* **17**, 20–2.

McNamara, R. S. (1977). Address to the Massachusetts Institute of Technology, 28 April. International Bank for Reconstruction and Development, 1818H Street, NW, Washington, DC 20433.

McNamara, R. S. (1982). The road ahead. Address to University of the Witwatersrand, Johannesburg, October 21.

Meadows, D. H., Randers, J., and Behrens, W. W. (1972). *The limits of growth. A report for the Club of Rome.* Pan Books, London, 205 pp.

Mikkelborg, E. (1974). Looking over the Black Angel's shoulder. *Can. Min. J.* August.

Minerals Exploration NWT (1976, 1977, 1978, 1979, 1980, 1981). Annual Reports, by Geology Office Staff, Dept. Indian and Northern Affairs, Yellowknife, Canada.

Mining Annual Review (1983). Gemstones, diamonds. *Min. Ann. Rev.* 119–21.

Mining Engineering (1984). US platinum, palladium lode development considered., *Min. Eng.* **36**, 11.

Mining Journal (1981). Black Angel reviewed. *Min. J.* May 29, 405.

Mining Journal (1982). Yes to UNCLOSS—regretfully. *Min. J.* **299**, 317–18.

Mining Journal (1983). Big fall in mined metal value. *Min. J.* **300**, 125

Mining Magazine (1982). Polaris Mine. *Mining Mag.* September, 180–93.

Mining Magazine (1983). Remote NW Territories gold mine. *Mining Mag.* February, 105.

Michalski, W. (1978). Industrial, raw materials; physical vs political economic and social scarcity of minerals. Interfutures project. *OECD Observer* **93**, 13–18.

Mitchell, B. (1981). Cracks in the ice. *Wilson Q.* Autumn, 69–84.

Mitchell, B. (1982). The management of Antarctic mineral resources. International Institute for Environment and Development (IIED), unpublished document, Washington DC, 285 pp.

Mitchell, B. (1983). *Frozen stakes: the future of Antarctic minerals.* Earthscan Publication, House of Print, London, 135 pp.

Mitchell, B. and Tinker, J. (1979). *Antarctica and its resources.* Earthscan (International Institute for Environment and Development), House of Print, London, 98 pp.

Mosaic (1978). Mineral deposits of Gondwanaland. *Mosaic* **9**, September/October.

Mohide, T. P. (1979). PGM—Ontario and the World. Ontario Ministry for Natural Resources, Mining Policy background paper No. 7.

Naldrett, A. J. (1978). Partitioning of Fe, Co, Ni and Cu between sulphide liquid and basaltic melts and the composition of Ni–Cu sulphide deposits: a discussion. *Econ. Geol.* **73**, 1520–8.

Naldrett, A. J. (1981). Platinum Group Element deposits. In: Platinum Group Elements: mineralogy, geology, recovery (ed. L. J. Cabri), pp. 197–231. Can. Institute of Mining and

Metallurgy, Special Volume No. 23.

Naldrett, A. J. and Cabri, L. J. (1976). Ultramafic and related rocks: their classification and genesis with special reference to the concentration of nickel sulphides and PGM. *Econ. Geol.* **71**, 1131–158.

Naldrett, A. J., Hoffman, E. L., Green, A. H., Chou, C-L., Naldrett, S. R., and Alcock, R. A. (1979). The composition of Ni-sulphide ores with particular reference to their content of PGE and Au. *Can. Mineral.* **17**, 403–16.

Nature (1982). Going for good. India Antarctic Expedition. *Nature* **295,** 64.

Nature (1983*a*). What to do about Antarctica. *Nature* **301**, 551–2.

Nature (1983*b*). Soviet Antarctica. When will the miners move south? *Nature* **301**, 457.

Nature (1983*c*). India in Antarctica. Science—and politics—on ice. *Nature* **306**, 106–7.

Nature (1984). Antarctic mining regime at risk. *Nature* **307**, 105–6.

Naval Nuclear Power Unit, Fort Belvoir, VA (1973). Removal plan for the PM-3A nuclear power plant, McMurdo Station, Antarctica (unpublished report).

Neethling, D. (1983). Minerals and energy. In: *South Africa in the world economy* (ed. J. Matthews), pp. 25–51. McGraw-Hill, Johannesburg.

Neider, C. (1980). *Beyond Cape Horn. Travels in the Antarctic.* Sierra Club Books, San Francisco, 387 pp.

Neuberg, H. A. C., Thiel, E., Walker, P. T., Behrendt, J. C., and Aughenbaugh, N. B. (1959). The Filchner Ice Shelf. *Ann. Ass. Am. Geogr.* **49**, 110–19.

Newman, S. C. (1973). Platinum. *Trans. Inst. Min. Metall.* **82**, A52–A68.

New Scientist (1982). Antarctic Resources beyond the Falklands. *New Sci.* 27 May, 561–3.

New Scientist (1983). Foreign office to sit on research council. *New Sci.* January, **97**, 217.

Nicolaysen, L. O., Day, P. W., and Hoch, A. (1982). Loss of deep mantle carbonic reduced volatiles during N-polarity, storage of these volatiles during R-polarity. In: Conference on planetary volatiles (ed. R. O. Pepin and R. O'Connel), pp. 119–21. Lunar and Planetary Institute Report No. 83–01.

Nicolaysen, L. O. (1983). On episodic, globally synchronous discharge of deep mantle volatiles at hotspots and its physical basis. IUGG inter-disciplinary Symposia, Hamburg, August 1983, **1**, p. 294.

Nordhaus, W. (1973). The allocation of energy resources. *Brookings papers on Economic Activity* **3**, 529–70.

Northern Natural Resource Development (1981). Requirements, procedures and legislation. Indian and Northern Affairs Canada, Ottawa, QS 8294-000-ER A1, 31 pp.

OECD (1979). Interfutures. facing the Future, mastering the probable and managing the unpredictable, OECD, Paris. 425 pp.

O'Hara, T. A. (1980). Quick guides to the evaluation of ore bodies. *Can. Inst. Min. Bull.* February, 87–99.

Olympic Dam Project (1982). A short summary of the environmental impact statement. Roxby Management Services (Pty) Ltd., Unley, Australia, 15 pp.

Oppenheimer, H. F. (1984). De Beers Consolidated Mines Ltd; 1983 report and chairman's statement.

Page, N. J. (1977). Stillwater complex, Montana: succession, metamorphism and structure of the complex and adjacent rocks. US Geological Survey Prof. paper No. 999.

Page, N. J. (1979). Stillwater complex, Montana: structure, mineralogy and petrology of the basal zone, with emphasis on the occurrence of sulphides. US Geological Survey Prof. paper No. 1038.

Page, N. J., Rowe, J. J., and Hoffty, J. (1976). Platinum metals in the Stillwater complex, Montana. *Econ. Geol.* **71**, 1352–63.

Palmer, K. (1980). Mineral taxation policies in developing countries: an application of resource rent tax. International Monetary Fund Staff Papers, Vol. 27, pp. 517–42.

Parker, B. C., ed. (1978). Environmental impact in Antarctica. Selected papers by scientists addressing impact assessment, monitoring and potential impact on man's activity in the Antarctic. Virginia Polytechnic and State University, Blacksburg, VA, 390 pp.

Pelissonier, H. and Michel, H. (1972). Les dimensions des gisements des cuirre du Monde. Essai de metallogenie quantitative. Bureau du Recherches Geologiques et Minieres, Memoir No. 57, 405 pp.

Peterson, G. (1979). Manganese nodules; future source of raw material; supplies and profitability. In: *Proceedings of the Second International Symposium, Hanover* 1979 (ed. F. Bender) pp. 69–78. Ed. Schweizerbart'sche, Stuttgart.

Philpott, J. E. (1983). Growth in fuel cell technology. A report of the United States national seminar. *Platinum Met. Rev.* **27**, 68–71.

Pinto, M. C. W. (1978). The international community and Antarctica. *Univ. Miami Law Rev.* **33**, 475–87.

Pinochet de la Barra, O. (1982). Politica para loss recursos antarticos. *Bol. Antarctica Chil.* **2**, 46–7.

Platinum Metals review (1983*a*). Application of the platinum metals. *Platinum Met. Rev.* **27**, 9.

Platinum Metals Review (1983*b*). Multi-megawatt fuel cell produces electricity: platinum catalysts promote efficient energy conversion. *Platinum Met. Rev.* **27**, 107.

Platinum Metals Review (1984). The United States national fuel cell seminar. *Platinum Met. Rev.* **28**, 19.

Pontecorvo, G. (1982). The economics of the mineral resources of Antarctica. In: *The new nationalism and the use of common spaces* (ed. Jonathon I. Charney). Allanheld, Osman and Co., Montclair, NJ.

Powell, D. L. (1983). Scientific and economic considerations relating to the conservation of marine living resources in Antarctica. In: *Antarctic resources policy: scientific, legal and political issues.* (ed. Vicuña, F. O.), pp. 111–18. Cambridge University Press, Cambridge, UK.

Pretorius, D. A. (1979). Foreword in: the Global status of South African minerals economy and data summaries of its key comodities by C. F. Vermaak. Geological Society of South Africa review paper No. 1, 57 pp.

Pride, D., and Moody, S. (1982). Trace element chemistry of mineralised rocks, Livingston Island (South Shetlands), Gerlache Strait and Southern Anvers Island. *Antarctic J. US* **17**, 43–5.

Quartino, B. J. and Rinaldi, C. A. (1976). La continuacion Antarctica de la faja cupro plumbifera de la Cordillera Patagonica y su relacion con los factores geologicas de localizacion. II Congress Ubero-Americano de Geologia Economica, Buenos Aires, Vol. 3, pp. 7–20.

Quigg, P. W. (1983). *A pole apart: the emerging issue of Antarctica.* McGraw-Hill, New York, 299 pp.

Rajamani, V., and Naldrett, A. J. (1978). Partitioning of Fe, Co, Ni and Cu between sulphide liquid and basaltic melts and the composition of Ni–Cu sulphide deposits. *Econ. Geol.* **73**, 82–93.

Ralling, C. (1983). Shackleton, BBC Publications, Jolly & Barber, UK, 263 pp.

Ravich, M. G., Fedorov, L. A., and Tarntin, O. A. (1982). Iron deposits of the Prince Charles Mountains. In: *Antarctic geosciences: geology and geophysics* (ed. C. Craddock). IUGS series B, No. 4, pp. 853–8. University of Wisconsin Press, Madison, WI.

Rand Daily Mail (1983). Europe may lift platinum sales. *Rand Daily Mail* 25.07.83, and: Platinum cloudy in short term. *Rand Daily Mail* 29.07.83, Johannesburg.

Reich, R. J. (1980). The development of Antarctic tourism. *Polar Rec.* **20**(126), 203–14.

Relph–Knight, L. (1982). Systems for Antarctica. *Build. Des.* March 5, 14–16.

Richardson, . (1982). It's an epic day for Canadian mining. *The Province,* Vancouver, B.C., Sunday, 1 August 1982.

Rich, R. (1982). A minerals regime for Antarctica. *Int. Comp. Law Q.* **31**, 709–25.

Ripley, E. A., Redmann, R. E., and Maxwell, J. (1978). Environmental impact of mining in Canada. Centre for Resource Studies, Queen's University, Kingston, Ontario. The National Impact of Mining Series, Vol. 7, 274 pp.

Roberts, B. (1977). Conservation in the Antarctic. *Phil. Trans. R. Soc. Lond.* **B97**, 97–104.

Roberts, N. (1983). New Zealand's interest in Antarctica. In: The Antarctic, preferred futures, constraints and choices. Proceedings of a seminar at Wellington, June 1983. New Zealand Institute of International Affairs, Wellington, Pamphlet No. 44, pp. 5–11.

Robin, G. de Q., Doake, C. S. M., Kohnen, H., Crabtree, R. D., Jordan, S. R., and Moller, D. (1983). Regime of the Filchner–Ronne Ice shelves, Antarctica. *Nature* **302**, 582–6.

Robson, A. (1979). Sequential exploitation of uncertain deposits of a depletable natural resource. *J. Econ. Theory* **21**, 88–110.

Roots, E. F. (1983). Resource development in polar regions: comments on technology. In: *Antarctic resources policy. Scientific, legal and political issues* (ed. F. O. Vicuña), pp. 297–315. Cambridge University Press, Cambridge, UK.

Rowley, P. D. and Pride, D. E. (1982). Metallic mineral resources of the Antarctic Peninsula. In: *Antarctic geosciences, geology and geophysics* (ed. Campbell Craddock), pp. 859–70. University of Wisconsin Press,

Madison, WI.

Rowley, P. D. and Williams, P. L. (1979)- Metallic mineral resources of Antarctica—development far in future.*Am. Ass. Pet. Geol. Bull.* **63**, 518 (abstract).

Rowley, P. D., Williams, P. L., and Pride, D. E. (1984). Metallic and nonmetallic mineral resources of Antarctica. In: *Mineral Resource Potential of Antarctica* (ed. J. F. Splettstoesser). University of Texas Press, Austin, TX (in press). Also in Behrendt (1983).

Rowley, P. D., Williams, P. L., and Schmidt, D. L. (1977). Geology of an Upper Cretaceous copper deposit in the Andean Province, Lassiter Coast, Antarctic Peninsula. US Geological Survey. Prof. Paper No. 984, 36 pp.

Rowley, P. D., Williams, P. L., Schmidt, D. L., Reynolds, R. L., Ford, A. B., Clark, A. H., Farrar, E., and McBridge, S. L. (1975). Copper mineralization along the Lassiter Coast of the Antarctic Peninsula. *Econ. Geol.* **70**, 982–92.

Runnels, D. D. (1970). Continental drift and economic minerals in Antarctica. *Earth Planet. Sci. Letters* **8**, 400–2.

Ryan, P. J. (1982). The geology of the Broken Hill ore deposit, Aggeneys, South Africa. *Proceedings of the 12th CMMI Congress* (ed. H. W. Glenn) pp. 181–92. South African Institute of Mining and Metallurgy, Johannesburg.

SAIRR (1983). *Survey of race relations in South Africa.* South African Institute of Race relations, Johannesburg. The Natal Witness Ltd. press, Pietermaritzburg, 650 pp.

SCAR (1977). Scientific Committee on Antarctic Research. Report of the group of experts on mineral exploration and exploitation. Annex 5 to Report of the 9th Consultative meeting, Antarctic Treaty, London, pp. 56–73. (Also known as the Holdgate Report.)

SCAR (1981). Scientific Committee on Antarctic Research. Constitution procedures and Structure. International Council of Scientific Unions, Cambridge, UK, 19 pp. (revised October 1981).

SCAR (1983). Scientific Committee on Antarctic Research. Bulletin No. 73, Scott Polar Research Institute, Cambridge, UK.

Schachter, O. (1977). *Sharing the world's resources.* Columbia University Press, New York. 172 pp.

Schatz, G, S, (1983). Sea floor engineering studies. *Geotimes* May, 14–15.

Schofield, E. A. (1976). Antarctica up for grabs. *Sierra Club Bull.* **62**, 17–19.

Schmidt, D. L. and Ford, A. B. (1969). *Geology of the Pensacola and Thiel Mountains.* Antarctic map folio series, Folio 12, Sheet 5. American Geographical Society Publication.

Schnellman, G. A. (1955). Mineral exploration. *Min. J. Lond. Ann. Rev.* pp. 79–85.

Scott, Capt. (1964). *R. F. Scott's last expedition: the personal journals of Captain R. F. Scott CVO RN on his journey to the south Pole.* The Folio Society, London. (First published in 1913 in two volumes.)

Sebinius, J. K. (1982). Financial aspects of Antarctic mineral regimes. Appendix in Mitchell (1982).

Shackleton, E. H. (1919). *South.* Heinemann, London.

Sharpe, M. R. (1982). Noble metals in the marginal rocks of the Bushveld complex. *Econ. Geol.* **77**, 1285–95.

Simmonds, K. R. (1983–). New directions in the Law of the Sea. Oceana Publications, London.

Singer, D. A. and Mosier, D. L. (1981). A review of regional mineral resource assessment methods. *Econ. Geol.* **81**, 1006–15.

Singer, D. A. and Overshine, A. T. (1979). Assessing metallic resources in Alaska. *Am. Sci.* **67**, 582–9.

Skinner, B. J. (1976). A second iron age ahead? *Am. Sci.* **64**, *258–69.*

Skinner, B. J. (1979). The frequency of mineral deposits. Alex L. du Toit Memorial lecture 16, *Geological Society of South Africa.* Annexure to Vol. 82, p. 12.

Skinner, B. J. and Peck, D. L. (1969). An immiscible sulphide melt from Hawaii. *Econ. Geol. Mon.* **4**, 310–22.

Slevich, S. (1968). *The ice continent today and tomorrow.* Leningrad, 280 pp. (with English summary).

Slevich, S. (1973). *Osnovnyye problemy osvoenia Antarkhiki* (Basic problems of Antarctic exploitation) Leningrad, 30 pp. Translation from joint publication research service, Arlington, VA, USA (1974).

Smith, A. G. and Hallam, A. (1970). The fit of the southern continents. *Nature* **225**, 139–44.

Smith, A. G., Hurley, A. M., and Briden, J. C. (1981). *Phanerozoic paleocontinental world maps.* Cambridge University Press, Cambridge, UK. 102 pp.

Smith, V. K. (1979). *Scarcity and growth recon-*

sidered. Resources for the future. Johns Hopkins University Press, Baltimore, 298 pp.

Sohn, K. H. (1979). Investments in the mineral-resources sector—opportunities and risks. In: *The mineral resource potential of the earth. Proceedings of the Second International Symposium, Hannover, 1979* (ed. F. Bender) pp. 132–48. Ed. Schweizerbart'sche, Stuttgart.

Sollie, F. (1983). Jurisdictional problems in relation to Antarctic mineral resources in political perspective. In: *Antarctic resources policy. Scientific, legal and political issues* (ed. F. O. Vicuña) pp. 317–35. Cambridge University Press, Cambridge, UK.

Splettstoesser, J. F. (1976). Mining in Antarctica. Survey of mineral resources and possible exploitation methods. In: *Proceedings of the Third International Conference on Port and Ocean Engineering under Arctic conditions,* Vol. 2, pp.1137–55.

Splettstoesser, J. F. (1978). Offshore development for oil and gas in Antarctica. In: *Proceedings of the Fourth International Conference on Port and Ocean Engineering under Arctic conditions,* pp. 811–20.

Splettstoesser, J. F. (1979). Underground technology for offshore hydrocarbon development in Antarctica. In: *Proceedings of the Fifth International Conference on Port and Ocean Engineering under Arctic conditions,* Vol. 3, pp. 233–45.

Splettstoesser, J. F. (1980). Coal in Antarctica. *Econ. Geol.* **75**, 936–42.

Splettstoesser, J. F. (1983). Mineral resources potential in Antarctica—review and predictions. In: *Antarctic earth science* (ed. R. L. Oliver, P. R. James, and J. B. Jago) p. 413. Australian Academy of Science, Canberra.

Splettstoesser, J. F. (1984). Coal potential in Antarctica. In: *Mineral resource potential of Antarctica* (ed. J. F. Splettstoesser). University of Texas Press, Austin, TX (in press).

Splettstoesser, J. F., Webers, G. F., and Waldrip, D. B. (1982). Logistic aspects of geological studies in the Ellsworth Mountains, Antarctica, 1979–1980. *Polar. Rec.* **21**, 147–59.

Stump, E., Self, S., Smit, J. H., and Colbert, P. V. (1981). Geological investigations in the La force Mountains and central Scott Glacier area. *Antarctic J. U.S.* **16,** 55–7.

Thompson, J. O. (1982). The international diamond industry. In: Proceedings of the 12th CMMI Congress (ed. H. W. Glen) pp. 21–5. South African Institute of Mining and Metallergy, Johannesburg.

Tinbergen, J. (1976). *Reshaping the international order (RIO) A report to the Club of Rome* (Tinbergen, co-ordinator). Signet, New York, 432 pp.

Tinker, J. (1979). Antarctica: towards a new internationalism. *New Sci.* 13 September, **7**, 99.

Tischler, S. E., Cawthorn, R. G., Kingston, G. A., and Maske, S. (1981). Magmatic Cu–Ni–PGE mineralisation at Waterfall Gorge, Insizwa, Pondoland, Transkei. *Can. Mineral.* **19**, 607–18.

Todd, E. P. (1983). NSF issues a directive on pollution control provisions of the Antarctic Conservation Act: a code of conduct for Antarctic expeditions and station activities. *Antarctic J. U.S.* **18**, 8–9.

Todd, S. G., Schissel, D. J., and Irvine, T. N. (1979). Lithographic variations associated with the platinum-rich zone of the Stillwater Complex. *Yb. Carnegie Inst. Wash.* **78**, 461–468.

Tonge, D. (1983). Third World eyes Antarctic resources; a report on a possible threat to the unique co-operation of a cosy club. *Financial Times,* London, 26 August.

Truswell, E. M. (1983). Geological implications of recycled polynomorphs in continental shelf sediments around Antrctica. In: *Antarctic earth science* (ed. R. L. Oliver, P. R. James, and J. B. Jago) pp. 334–9. Australian Academy of Science, Canberra.

Tucker, A. (1983*a*). Keeping the goods on ice. A report on the Antarctic Treaty nations' discussions in Bonn (July 1983) concerning the establishment of a minerals regime. *The Guardian Weekly,* 17 July, p. 8.

Tucker, A. (1983*b*). Greenpeace attacks plan to exploit Antarctica. *Guardian* 15 June.

Ulph, A. M. and Folie, G. M. (1980). Exhaustible resources and cartels. An intertemporal Nash–Cournet model. *Can. J. Econ.* **13**, 645–58.

Union of International Associations, ed. (1984). *Yearbook of International Organisations,* 1983–1984 (20th edn). K. G. Saur, Munchen.

United Nations' General Assembly (1983). Question of Antarctica, 38th Session, 1st Committee, Agenda Item 140, A/C.1/38/L.80, 28 November 1983 (limited distribution) 26 pp.

USA (1983). National Report No. 24 to SCAR, Washington, DC.

USARP (1982). Facts about the United States Antarctic Research Program. Division of polar programs. NSF publication, Washington DC, 9 pp.

USARP (1983). Cost of United States Government Antarctic activities (1953–1984). 24 March, information sheet 10, 3 pp.

US Department of State (1982). Final environmental impact statement on the negotiation of an international regime for Antarctic mineral resources. 73 pp., with five appendices. (Initial preliminary draft circulated in 1979.)

USSR (1981). National Report 23 to SCAR.

USSR (1982). National Report 24 to SCAR.

US Senate Committee on Foreign Relations (1980). Imports of minerals from South Africa by the United States and the OECD countries. US Government Printing Office, Washington, 46 pp.

Vale, P. C. J. (1980). The Atlantic Nations and South Africa. Economic constraints and community fracture. Unpublished Ph.D., University of Leicester, 446 pp.

Van den Broeck, H. and Cameron, D. S. (1984). Fuel cells for vehicle propulsion. Platinum metal catalysis used in alkaline units. *Platinum Met. Rev.* **28**, 46–52.

Van der Essen, A. (1983). The application of the Law of the Sea to the Antarctic continent. In: *Antarctic resources policy. Scientific, legal and political issues* (ed. F. O. Vicuña) pp. 231–42. Cambridge University Press, Cambridge, UK.

van Onselen, C. (1982). Studies in the social and economic history of the Witwatersrand 1884–1914. Vol. 1, New Babylon. Ravan Press, Johannesburg, 213 pp.

van Rensburg, W. C. J., and Pretorius, D. A. (1977). South Africa's strategic minerals—pieces on a continental chessboard (ed. Helen Glenn). Valiant Publishers, Johannesburg, 116 pp.

Vestgron Mines Ltd. (1975, 1978–1981). Annual Reports.

Vicuña, F. O., ed. (1983). *Antarctic resources policy: scientific, legal and political issues,* pp. 1–10 and pp. 243–51. Cambridge University Press, Cambridge, UK.

Vieira, C., Alarcón, B., Ambrus, J., Olcay, L. (1982). Metallic Mineralisation in the Gerlache Strait Region, Antarctica. In: *Antarctic geosciences; geology and geophysics* (ed. Campbell Craddock) pp. 871–83. IUGS series B, No. 4, University of Wisconsin Press, Madison.

von Gruenewaldt, G. (1977). The mineral resources of the Bushveld complex. *Minerals Sci. Eng.* **9**, 83–95.

Wade, A. (1940). The geology of the Antarctic continent and its relationship to neighbouring land areas. *Proc. R. Soc. Queensland* **52**, 1–12.

Waddel, C. H. (1982). The platinum industry. Proceedings of the 12th CMMI Congress (ed. H. W. Glen) pp. 43–4. South African Institute of Mining and Metallurgy, Johannesburg.

Walker, P. T. (1961). Study of some rocks and minerals from the Dufek Massif, Antarctica: International Geophysical Year World data centre A, Glaciology Report, No. 4, pp. 195–213.

Ward, B., and Dubois, R. (1972). *Only one earth: the care and maintenance of a small planet.* Penguin Books, Harmondsworth, UK. 304 pp.

Washburn, A. L. (1980). Focus on Polar research. *Science* **209**, 643–53.

Watson, J. V. (1978). Ore-deposition through geological time. *Proc. R. Soc. Lond.* **A362**, 305–28.

Watson, J. V. (1980). Metallogenesis in relation to mantle heterogeneity. *Phil. Trans. R. Soc. Lond. A297*, 347–52.

Western Mining (1976). The Black Angel. Canadian mining experience in the high Arctic. *West. Min.* October, 21–4.

Wilson, J. T. (1961). *I.G.Y. The year of the new moon.* Michael Joseph, London, 352 pp.

Wojciechowski, M. J. (1981). Eastern Arctic study. Case study series: the Nanisivik mine. Queens University Centre for Resource Studies, Kingston, Canada, 71 pp.

World Bank (1982*a*). Somalia; proposed petroleum contract fiscal terms. Unpublished manuscript, Washington DC.

World Bank (1982*b*). 1981 *World Bank atlas: gross national product, population and growth rates.* IBRD, Washington DC, 24 pp.

Wright, N. A., and Williams, P. L. (1974). Mineral resources of Antarctica. US Geological Survey circular No. 705, 29 pp.

Zeller, E. J., and Dreschhoff, G. A. M. (1983). Uranium resource evaluation in Antarctica. In: *Mineral resource potential of Antarctica* (ed. J. Splettstoesser) University of Texas Press, Austin, TX (in press).

Zumberge, J. H. (1979*a*). Mineral resources and geopolitics in Antarctica. *Am. Sci.* **67**, 68–77.

Zumberge, J. H., ed. (1979*b*). Possible environmental effects of mineral exploration in Antarctica. EAMREA Report 1979.

Zumberge, J. H. (1981). Potential mineral resource availability and possible environment problems in Antarctica. In: *New nationalism and the use of common spaces.* (ed. J. I. Charney) Allandheld, Osman and Co., Montclair, NJ.

Zwally, H. J., Comiso, J., Parkinson, C. L., Campbell, W. J., Carsey, F. D., and Gloersen, P. (1983). Antarctic Sea Ice 1973–1976. Satellite passive-microwave observations. NASA, Washington DC, special publication No. 459, 206 pp.

Zwartendyk, J. (1981). Economic Issues: mineral resource adequacy in the long-term supply of minerals. *Econ. Geol.* **76**, 999–1005.

INDEX